Essentials of Toxicology

Essentials of Toxicology

Jaenette Mountford

SYRAWOOD
PUBLISHING HOUSE

New York

Published by Syrawood Publishing House,
750 Third Avenue, 9th Floor,
New York, NY 10017, USA
www.syrawoodpublishinghouse.com

Essentials of Toxicology
Jaenette Mountford

International Standard Book Number: 978-1-68286-826-3 (Hardback)

Cataloging-in-Publication Data

Essentials of toxicology / Jaenette Mountford.
    p. cm.
Includes bibliographical references and index.
ISBN 978-1-68286-826-3
1. Toxicology. 2. Poisoning. 3. Poisons. I. Mountford, Jaenette.
RA1211 .E87 2019
615.9--dc23

TABLE OF CONTENTS

Permissions

Index

PREFACE

The study of the negative effects of chemical substances on living organisms are studied under toxicology. The field also encompasses the diagnosis and treatment of exposures of organisms to toxins and toxicants, relationship of the dose with its effect as well as the factors influencing toxicity of the chemical agent. Adverse reactions generally depend on the route of exposure, dosage and factors of age, sex, health, species, etc. Toxicity experiments can be conducted in vivo, in vitro or in silico. This book aims to shed light on some of the unexplored aspects of toxicology. It elucidates new techniques of toxicology testing and their applications in a multidisciplinary approach. In this textbook, constant effort has been made to make the understanding of the difficult concepts of toxicology as easy and informative as possible, for the readers.

A detailed account of the significant topics covered in this book is provided below:

Chapter 1- The study of the negative impacts of chemical substances on living organisms is under the domain of toxicology. It also involves the diagnosis and treatment of toxic exposures. This chapter has been carefully written to provide an introduction to toxicology and discusses the fundamentals of toxin, toxicant and poison.

Chapter 2- Toxicology is an important science with three primary branches, medical, clinical and computational toxicology. Depending on the area of application of this science, it can be further divided into many other sub-disciplines, such as environmental toxicology, aquatic toxicology, ecotoxicology, medical toxicology, etc. which have been discussed in extensive detail in this chapter.

Chapter 3- A toxin is a poisonous substance that is produced within the cells of living organisms. It can be peptides or small molecules, and vary in toxicity from minor complications to death. There are different classifications of toxins, such as microbial toxins, neurotoxins, plant and animal toxins, which have been discussed in comprehensive detail in this chapter.

Chapter 4- Toxicological testing is conducted in order to determine how damaging a substance is to a living or non-living organism. It is essential for medicine and pesticide testing. The aim of this chapter is to investigate and analyze the varied tests for toxicity, such as in vivo and vitro testing, Draize test, Reinsch test, etc.

Chapter 5- The degree to which a chemical substance can damage a living organism is known as its toxicity. It is influenced by a variety of factors, such as the pathway of administration, the duration of exposure, the number of exposures, etc. This chapter discusses the diverse aspects of toxicity, its symptoms and manifestations, through an analysis of acute, subchronic, chronic and developmental toxicity.

Chapter 6- Exposure to a toxic substance can affect the health of an organism. This chapter closely examines the toxic effects of substances such as arsenic, carbon monoxide, chloride gas, fluoride, sulphuric acid, etc.

Chapter 7- Specific-target organ toxins damage specific organs only. Toxins can target the respiratory system, reproduction system, neurological system, cardiovascular system, etc. The various toxic effects of organ specific toxins have been discussed in this chapter.

Chapter 8- The complete understanding of toxicity requires a study of acute toxicity, toxicity class, lethal dose, fixed dose procedure, bioaccumulation, etc. These have been extensively elaborated in this chapter.

It gives me an immense pleasure to thank our entire team for their efforts. Finally in the end, I would like to thank my family and colleagues who have been a great source of inspiration and support.

Jaenette Mountford

Chapter 1
An Introduction to Toxicology

The study of the negative impacts of chemical substances on living organisms is under the domain of toxicology. It also involves the diagnosis and treatment of toxic exposures. This chapter has been carefully written to provide an introduction to toxicology and discusses the fundamentals of toxin, toxicant and poison.

Toxicology is the study of chemicals that can cause problems for living things. It's a wide-ranging field: The chemicals may occur naturally or have been created in a lab or factory; the living things may be humans, pets, livestock, or microbes living in a pond. And the problems that concern toxicologists stretch from inconvenience to disaster— from mild skin irritation, for instance, to death.

Toxicology has a broad scope. It deals with toxicity and mechanisms of toxicity of chemicals used in medicine for diagnostic, preventive and therapeutic purposes, in the food industry as direct and indirect additives, in agriculture as pesticides, growth regulators, artificial pollinators, and in the chemical industry as solvents, components, and intermediates of plastics and many other types of chemicals. It also concerned with the health effects of metals, petroleum products, wastes of paper and pulp industry, air pollutants, and animal and plant toxins. Risk assessment of deleterious of health effects induced by chemicals is a major enterprise in toxicology.

Many of the problems in toxicology require an integrative approach to problem solving. Therefore toxicologists utilise variety of approaches that span several sectors of science to obtain a comprehensive picture of the risks associated with chemicals under investigation. These sectors include e.g. human and animal anatomy and physiology, chemistry, biochemistry, pharmacology, biology and social sciences. A collaboration between experts of different sectors is crucial to gain a comprehensive picture of the risks induced by chemicals. In addition it is important to be able to communicate the risks identified in an understandable, clear manner. Broad know-how and comprehensive ability to understand the full picture is required for the successful toxicological risk assessment, in the communication of the conclusions and in the decision making process.

Toxicology is divided to multiple special expertise areas such as regulatory, mechanistical or descriptive toxicology. On the other hand the areas can be specified according to the functions such as clinical toxicology, ecotoxicology, forensic toxicology or occupational toxicology. All these sectors of toxicology aim to make the world safer place for living.

Importance of Dose

The dose of the substance is an important factor in toxicology, as it has a significant relationship with the effects experienced by the individual. It is the primary means of classifying the toxicity of the chemical, as it measures the quantity of the chemical, or the exposure to the substance. All substances have the potential to be toxic if given to living organisms in the right conditions and dose.

LD_{50} is a common term used in toxicology, which refers to the dose of a substance that displays toxicity in that it kills 50% of a test population. In scientific research, rats or other surrogates are usually used to determine toxicity and the data are extrapolated to use by humans.

A conventional relationship between dose and toxicity has traditionally been accepted, in that greater exposure to a chemical leads to higher risk of toxicity. However, this concept has been challenged by a study of endocrine disruptors and may not be a straightforward relationship.

Chemical Toxicology

Chemical toxicology is a subspecialty of toxicology that focuses on the structure of chemical agents and how it affects their mechanism of action on living organisms.

It is a multidisciplinary field that includes computational and synthetic chemistry, in addition to people who specialize in the fields of proteomics, metabolomics, drug discovery, drug metabolism, bioinformatics, analytical chemistry, biological chemistry and molecular epidemiology. It relies on technological advances to help understand the chemical components of toxicology more comprehensively.

Toxicology and Pharmacology Differences

Toxicology and pharmacology are both studies that involve an understanding of chemical properties and their actions on the body, but differ considerably in other areas.

Pharmacology primarily focuses on the therapeutic effects of pharmaceutical substances and how they can be used most effectively for medical purpose. On the contrary, toxicology is more closely related to the adverse effects that can occur in living organisms that come into contact with chemicals. Toxicologists are also more concerned with measuring the risk of certain substances with risk assessment tools.

Toxicologist

A toxicologist is someone who has studied toxicology and works with materials and chemicals to determine the toxic effects they may have on the environments and living organisms. People who are methodical and scientific are well suited for a career in toxicology.

Testing Methods

Toxicity experiments may be conducted *in vivo* (using the whole animal) or *in vitro* (testing on isolated cells or tissues), or *in silico* (in a computer simulation).

Non-human Animals

The classic experimental tool of toxicology is testing on non-human animals. An example of a model organism is *Galleria mellonella,* which can replace small mammals to study toxicology in vivo. As of 2014, such animal testing provides information that is not available by other means about how substances function in a living organism.

Alternative Testing Methods

While testing in animal models remains as a method of estimating human effects, there are both ethical and technical concerns with animal testing.

Since the late 1950s, the field of toxicology has sought to reduce or eliminate animal testing under the rubric of "Three Rs" - reduce the number of experiments with animals to the minimum necessary; refine experiments to cause less suffering, and replace *in vivo* experiments with other types, or use more simple forms of life when possible.

Computer modeling is an example of alternative testing methods; using computer models of chemicals and proteins, structure-activity relationships can be determined, and chemical structures that are likely to bind to, and interfere with, proteins with essential functions, can be identified. This work requires expert knowledge in molecular modeling and statistics together with expert judgment in chemistry, biology and toxicology.

In 2007 the American NGO National Academy of Sciences published a report called "Toxicity Testing in the 21st Century: A Vision and a Strategy" which opened with a statement: "Change often involves a pivotal event that builds on previous history and opens the door to a new era. Pivotal events in science include the discovery of penicillin, the elucidation of the DNA double helix, and the development of computers. ...Toxicity testing is approaching such a scientific pivot point. It is poised to take advantage of the revolutions in biology and biotechnology. Advances in toxicogenomics, bioinformatics, systems biology, epigenetics, and computational toxicology could transform toxicity testing from a system based on whole-animal testing to one founded primarily on in vitro methods that evaluate changes in biologic processes using cells, cell lines, or cellular components, preferably of human origin." As of 2014 that vision was still unrealized.

In some cases shifts away from animal studies has been mandated by law or regulation; the European Union (EU) prohibited use of animal testing for cosmetics in 2013.

Types

Medical Toxicology

Medical toxicology is the discipline that requires physician status (MD or DO degree plus specialty education and experience).

Clinical Toxicology

Clinical toxicology is the discipline that can be practiced not only by physicians but also other health professionals with a master's degree in clinical toxicology: physician extenders (physician assistants, nurse practitioners), nurses, pharmacists, and allied health professionals.

Computational Toxicology

Computational toxicology is a discipline that develops mathematical and computer-based models to better understand and predict adverse health effects caused by chemicals, such as environmental pollutants and pharmaceuticals. Within the *Toxicology in the 21st Century* project, the best predictive models were identified to be Deep Neural Networks, Random Forest, and Support Vector Machines, which can reach the performance of in vitro experiments.

Toxicant

Toxicant is usually used when referring to a toxic agent that is produced by, or a by-product of, man-made activities.

Toxic agents are classified in a number of ways depending on the interests and needs of the classifier. Dioscorides classified substances using general characteristics i.e., whether they are toxic or therapeutic. An early scheme by Orfila classified substances as being of animal, vegetable or mineral origin. No single classification is applicable for the entire spectrum of toxic agents and combinations of classification systems based on other factors may be needed to provide the best rating system. Nevertheless, classification systems that take into account both the chemical and biological properties of the agent, and the exposure characteristics are most likely to be useful for toxicology in general. Many classification schemes for toxic agents are available which are based on following points:

Based on the source of toxicants

a. Plant toxicants, e.g., morphine, curare, strychnine.

b. Animal toxicants, e.g. toxins (zootoxins), venoms.

c. Mineral toxicants, e.g. copper, lead, selenium, iron.

d. Synthetic toxicants, e.g. organophosphates, carbamates, aluminium phosphide.

Based on the Physical state of toxicants

a. Gaseous toxicants, e.g. hydrocyanic acid (HCN), sulphur dioxide, carbon monoxide, phosphine.

b. Liquid toxicants, e.g. sulphuric acid, carbon disulphide, nicotine.

c. Solid toxicants, e.g. strychnine, opium, atropine.

d. Dust toxicants, e.g. asbestos dust, silicon dust, metallic dusts.

Based on the physical characteristics

a. Inflammable/Non-inflammable,

b. Explosive/Non-explosive

Based on the physical effects

a. Irritant/Non-irritant

b. Corrosive/Non-corrosive

Based on the Target organ/system: It is difficult to classify a toxicant on the basis of its target organ or system as it may affect other systems also. However, action on the primary site has been taken as a basis of classification.

a. Hepatotoxins, e.g. carbon tetrachloride, aflatoxins, phenol.

b. Neurotoxins, e.g. organophosphorus insecticides, pyrethroids, anaesthetics, nicotine.

c. Nephrotoxins, e.g. heavy metals (lead, arsenic, cadmium), oxalates.

d. Pulmonotoxicants, e.g. alpha-naphthylthiourea (ANTU), hydrogen sulphide, ammonia gas.

e. Haematotoxins, e.g. warfarin, cyanide, phenothiazine, snake venom (pit viper venom).

   f.   Dermatotoxicants, e.g. coal tar compounds (petroleum oils), heavy metals (arsenic, mercury), p-tertiarybutyl phenol.

Based on the Chemical nature/structure of toxicants:

a. Inorganic toxicants: These include metals, metalloids, non-metals and their salts and derivatives, acids and alkalies.

1. Metals, e.g., lead, copper, mercury, antimony.

2. Non-metals, e.g. phosphorus, sulphur, chlorine, nitrate/nitrite.

3. Acids and alkalies, e.g. hydrochloric acid, sulphuric acid, potassium hydroxide.

b. Organic toxicants: These include all carbon compounds other than the oxides of carbon, the carbonates, and the metallic carbides and cyanides.

1. Hydrocarbons, e.g. cyclopropane, benzene, paraffin, naphthalene.

2. Halogen derivatives of hydrocarbons, e.g. chloroform, BHC, DOT, carbon tetrachloride.

3. Alcohols and phenols, e.g. methyl alcohol, ethyl alcohol, phenol, cresol, pentachlorophenol.

4. Ethers, e.g. diethyl ether, divinyl ether.

5. Aldehydes and ketones, e.g. formaldehyde, paraldehyde.

6. Organic acids, e.g. formic acid, phenoxy acetic acid, salicylic acid.

7. Esters, e.g. organophosphorus insecticides, succinylcholine.

8. Amines, e.g. adrenaline, ephedrine, amphetamine.

9. Amides, e.g. phenacetin, sulphonamides.

10. Glycosides, e.g. digitoxin, cyanogenetic glycosides.

11. Alkaloids, e.g. nicotine, atropine, strychnine.

12. Proteins, e.g. ricin, crotin, abrin.

Based on the Analytical behaviour of toxicants: Toxicants can be classified into separate characteristic groups according to the analytical procedures involved. In the Stas-Otto scheme, toxicants have been divided into the following groups.

   a.   Volatile toxicants, e.g. hydrocyanic acid, alcohols, acetone, phenol, chloral hydrate.

   b.   Extractive toxicants

      1.   Toxicants extractable by ether from acid solution, e.g. organic acids,

2. Toxicants extractable by ether from alkaline solution e.g. alkaloids

3. Metals and metalloids, e.g. Copper, mercury, zinc, silver, antimony.

Based on the Toxic effects: It is difficult to classify toxicants on the basis of toxic effects as a single compound may have number of effects.

a. Carcinogens, e.g. thiouracil, vinyl chloride, nickel.

b. Mutagens, e.g. ethyl methane sulphonate, UV light, nitrogen mustards, nitroso compounds.

c. Teratogens, e.g. phenylmercuric acetate, triazines, thalidomide.

d. Clastogens, e.g. UV light, caffeine.

Based on the principal uses of toxicants:

a. Insecticides, e.g. organophosphorus insecticides, carbamates, pyrethroids.

b. Fungicides, e.g. captan, folpet, pentachlorophenol.

c. Herbicides, e.g. triazine, paraquat, 2,4-D.

d. Rodenticides, e.g. warfarin, fluoroacetate, red squill.

e. Food additives:

1. Preservatives, e.g. ascorbic acid, sodium bisulphite.

2. Antioxidants, e.g. ascorbic acid.

3. Emulsifying agents, e.g. cholic acid, desoxycholic acid.

4. Colouring agents, e.g. amaranth, tartrazine.

5. Anticaking agents, e.g. aluminium-calcium silicate.

Based on the Toxicity/Poisoning potential:

	Group	Lethal dose
a.	Extremely toxic	< 1 mg/kg
b.	Highly toxic	1-50 mg/kg
c.	Moderately toxic	50-500 mg/kg
d.	Slightly toxic	0.5-5 g/kg
e.	Practically non-toxic	5-15 g/kg
f	Relatively harmless	> 15 g/kg.

Based on the Mechanism of action:

a. Anticholinesterase agents/cholinesterase inhibitors, e.g. organophosphorus in-secticides, carbamates.

b. Sulphydryl (-SH) inhibitors, e.g. mercury, arsenic.

c. Protoplasmic toxicants, e.g. heavy metals.

d. Corrosive toxicants, e.g. caustic alkalies, acids, heavy metals, irritant gases.

e. Methaemoglobin producers, e.g. nitrite

f. Inhibitors of mixed function oxidases (MFO), e.g. pipronyl butoxide.

g. Inhibitors of Kreb's cycle, e.g. fluoroacetate.

h. Uncoupler of oxidative phosphorylation, e.g. dinitrophenols, chlorophenol fun-gicides.

Based on the environmental and human health consideration:

• Air pollutants, water pollutants, radiation hazards, occupational hazards, etc.

Toxin

A toxin is a chemical substance which damages an organism. A toxin may be as simple as an ion or atom which negatively interferes with a cell. A toxin can also be in the form of complex molecules such as the proteins found in snake venom. Still other atoms and chemicals emit radiation, which has toxic effects on an organism. The effects of toxins vary widely in different organisms, and with different toxins. The end result of the strongest toxins is death, due to the damage they cause across the different cells of an organism. Different toxins act in different ways to affect the cells they damage.

Effects of a Toxin

The effects of a toxin are entirely determined by the biochemical reactions which take place when a potential toxin is introduced into an organism. Toxicologists must also take into consideration the environment in which the organisms lives. As mentioned before, things like pressure, heat, and metabolic rate can drastically change the effects of a toxin. Further, not all organisms react to toxins in the same way. Each organisms, even within a species, is essentially a unique biochemical factory. Some organisms are better equipped to handle certain toxins than others.

The specific effects of a toxin are determined by how it interacts with the cells of the organisms. Some toxins work by disrupting ion channels within the cells, while others can destroy the cell membrane or mutate the DNA. All of these conditions will eventually lead to the organism dying if the toxin is not removed. Organisms use their immune systems to target and remove protein-based toxins, while they rely on the filtration of their blood to remove ions and other free radicals. The damage done by a toxin is determined by its structure, atomically.

Toxin vs. Toxicant

Some areas of science prefer to define toxin as any harmful substance of purely biological origin. Anything produced artificially they refer to as a toxicant. However, other fields of science refer to pesticides like DDT as environmental toxins, and don't use the term "toxicant" to define the artificial toxins. Due to the nature of the field of Toxicology, and how it sees any substance as a toxin once it becomes harmful.

Poison

A poison is a substance capable of producing adverse effects on an individual under appropriate conditions. The term "substance" is almost always synonymous with "chemical" and includes drugs, vitamins, pesticides, pollutants, and proteins. Even radiation is a toxic substance. Though not usually considered to be a "chemical," most radiations are generated from radioisotopes, which are chemicals. The term "adverse effects" above refers to the injury, such as structural damage to tissues. "Appropriate conditions" refers to the dosage of the substance that is sufficient to cause these adverse effects. The dose concept is important because according to it even a substance as innocuous as water is poisonous if too much is ingested. Whether a drug acts as a therapy or as a poison depends on the dose.

Classification of a Poison

Poisons are of such diverse natures that they are classified by origin, physical form, chemical nature, chemical activity, target site, or use.

Classification Based on Origin

Poisons are of microbial, plant, animal, or synthetic origin. Microbial poisons are produced by microscopic organisms such as bacteria and fungi. Botulinus toxin, for example, is produced by the bacterium Clostridium botulinum and is capable of inducing weakness and paralysis when present in underprocessed, nonacidic canned foods or in other foods containing the spores. An example of a plant toxin is the belladonna alkaloid hyoscyamine, which is found in belladonna (Atropa belladonna) and jimson weed (Datura stramonium).

Animal poisons are usually transferred through the bites and stings of venomous terrestrial or marine animals, the former group including poisonous snakes, scorpions, spiders, and ants, and the latter group including sea snakes, stingrays, and jellyfish. Synthetic toxins are responsible for most poisonings. "Synthetic" refers to chemicals manufactured by chemists, such as drugs and pesticides, as well as chemicals purified from natural sources, such as metals from ores and solvents from petroleum. Synthetic toxins include pesticides, household cleaners, cosmetics, pharmaceuticals, and hydrocarbons.

Classification Based on Physical Form

The physical form of a chemical—solid, liquid, gas, vapor, or aerosol—influences the exposure and absorbability.

Because solids are generally not well absorbed into the blood, they must be dissolved in the aqueous liquid lining the intestinal tract if ingested or the respiratory tract if inhaled. Solids dissolve at different rates in fluids, however. For example, compared with lead sulphate granules, granules of lead are practically nontoxic when ingested, because elemental lead is essentially insoluble in water, while lead sulphate is slightly soluble and absorbable. Even different-sized granules of the same chemical can vary in their relative toxicities because of the differences in dissolution rates. For example, arsenic trioxide is more toxic in the form of smaller granules than is the same mass of larger granules because the smaller granules dissolve faster.

A poison in a liquid form can be absorbed by ingestion or by inhalation or through the skin. Poisons that are gases at room temperature (e.g., carbon monoxide) are absorbed mainly by inhalation, as are vapors, which are the gas phase of substances that are liquids at room temperature and atmospheric pressure (e.g., benzene). Because organic liquids are more volatile than inorganic liquids, inhalation of organic vapors is more common. Although vapors are generally absorbed in the lungs, some vapors that are highly soluble in lipids (e.g., furfural) are also absorbed through the skin.

Aerosols are solid or liquid particles small enough to remain suspended in air for a few minutes. Fibers and dust are solid aerosols. Aerosol exposures occur when aerosols are deposited on the skin or inhaled. Aerosol toxicity is usually higher in the lungs than on the skin. An example of a toxic fiber is asbestos, which can cause a rare form of lung cancer (mesothelioma).

Many liquid poisons can exist as liquid aerosols, although highly volatile liquids, such as benzene, seldom exist as aerosols. A moderately volatile liquid poison can exist as both an aerosol and as a vapor. Airborne liquid chemicals of low volatility exist only as aerosols.

Classification Based on Chemical Nature

Poisons can be classified according to whether the chemical is metallic versus nonmetallic, organic versus inorganic, or acidic versus alkaline. Metallic poisons are often eliminated from the body slowly and accumulate to a greater extent than nonmetallic poisons and thus are more likely to cause toxicity during chronic exposure. Organic chemicals are more soluble in lipids and therefore can usually pass through the lipid-rich cell membranes more readily than can inorganic chemicals. As a result, organic chemicals are generally absorbed more extensively than inorganic chemicals. Classification based on acidity is useful because, while both acids and alkalis are corrosive to the eyes, skin, and intestinal tract, alkalis generally penetrate the tissue more deeply than acids and tend to cause more severe tissue damage.

Classification Based on Chemical Activity

Electrophilic (electron-loving) chemicals attack the nucleophilic (nucleus-loving) sites of the cells' macromolecules, such as deoxyribonucleic acid (DNA), producing mutations, cancers, and malformations. Poisons also may be grouped according to their ability to mimic the structure of certain important molecules in the cell. They substitute for the cells' molecules in chemical reactions, disrupting important cellular functions. Methotrexate, for example, disrupts the synthesis of DNA and ribonucleic acid (RNA).

Other Classifications

Unlike the classifications described above, there is usually no predictive value in classification by target sites or by uses. Such classifications are done, however, to systematically categorize the numerous known poisons. Target sites include the nervous system, the cardiovascular system, the reproductive system, the immune system, and the lungs, liver, and kidneys. Poisons are classified by such uses as pesticides, household products, pharmaceuticals, organic solvents, drugs of abuse, or industrial chemicals.

References

- Hodgson, Ernest (2010). A Textbook of Modern Toxicology. John Wiley and Sons. p. 10. ISBN 978-0-470-46206-5.

- What-is-Toxicology, health: news-medical.net, Retrieved 11 May 2018

- Wennig, Robert (April 2009). "Back to the roots of modern analytical toxicology: Jean Servais Stas and the Bocarmé murder case". Drug Testing and Analysis. 1 (4): 153–155. doi:10.1002/dta.32. PMID 20355192.

- Ottoboni, M. Alice (1991). The dose makes the poison : a plain-language guide to toxicology (2nd ed.). New York, N.Y: Van Nostrand Reinhold. ISBN 978-0-442-00660-0

- Bruin, Yuri. et. al (2009). Testing methods and toxicity assessment (Including alternatives). Academic Press. pp. 497–514. doi:10.1016/B978-0-12-373593-5.00060-4. ISBN 9780123735935.

- Classification-of-toxicants: drjaggyvpt.blogspot.com, Retrieved 10 July 2018

- Bhat, Sathyanarayana; Udupa, Kumaraswamy (1 August 2013). "Taxonomical outlines of bio-diversity of Karnataka in a 14th century Kannada toxicology text Khagendra Mani Darpana". Asian Pacific Journal of Tropical Biomedicine. 3 (8): 668–672. doi:10.1016/S2221-1691(13)60134-3. PMC 3703563. PMID 23905027

Chapter 2

Prominent Branches of Toxicology

Toxicology is an important science with three primary branches, medical, clinical and computational toxicology. Depending on the area of application of this science, it can be further divided into many other sub-disciplines, such as environmental toxicology, aquatic toxicology, ecotoxicology, medical toxicology, etc. which have been discussed in extensive detail in this chapter.

Environmental Toxicology

Environmental toxicology is the field of study in the environmental sciences that is concerned with the assessment of toxic substances in the environment.

Although it is based on toxicology, environmental toxicology draws heavily on principles and techniques from other fields, including biochemistry, cell biology, developmental biology, and genetics. Among its primary interests are the assessment of toxic substances in the environment, the monitoring of environments for the presence of toxic substances, the effects of toxins on biotic and abiotic components of ecosystems, and the metabolism and biological and environmental fate of toxins.

Assessment and Monitoring of Toxic Substances

Toxins affect the environment and organisms in a variety of ways, from having little negative impact on certain abiotic factors or resistant organisms to killing animals and destroying major components of ecosystems. The extent of damage depends on the type and structure of the toxic substance; the age, the size, and the species of the organism; and the temperature and the physical and chemical characteristics of the environment (whether terrestrial or aquatic). Knowledge of how these factors interact is critical to understanding how best to prevent or reduce exposure or remove a toxin from the environment (environmental remediation).

The assessment of toxicity at the levels of whole organism, cell, and gene is one way by which researchers are able to determine how much of a toxin an organism can be exposed to before adverse effects set in. Different assays are used for toxicity assessment, including acute and subacute toxicity assays, sediment toxicity assays, and genotoxicity assays. The determination of safe exposure levels in animals plays a key role in the development of regulations that dictate how toxic substances are to be handled and disposed of. There are also methods by which scientists are able to estimate the quantity of a given toxic substance in the environment.

The identification of ways to monitor for chemicals in the environment is an important aspect of environmental toxicology. Monitoring typically is based on the detection of sensitive biochemical markers (e.g., proteins), the levels of which change in the presence of a given toxin, or on changes in individual "indicator" species, the well-being of which serves as a measure of environmental conditions and the health of other species.

Sources of Environmental Toxicity

There are many sources of environmental toxicity that can lead to the presence of toxicants in our food, water and air. These sources include organic and inorganic pollutants, pesticides and biological agents, all of which can have harmful effects on living organisms. There can be so called point sources of pollution, for instance the drains from a specific factory but also non-point sources (diffuse sources) like the rubber from car tires that contain numerous chemicals and heavy metals that are spread in the environment.

PCBs

Polychlorinated biphenyls (PCBs) are organic pollutants that are still present in our environment today, despite being banned in many countries, including the United States and Canada. Due to the persistent nature of PCBs in aquatic ecosystems, many aquatic species contain high levels of this chemical. For example, wild salmon (*Salmo salar*) in the Baltic Sea have been shown to have significantly higher PCB levels than farmed salmon as the wild fish live in a heavily contaminated environment.

Heavy Metals

Heavy metals found in food sources, such as fish can also have harmful effects. These metals can include mercury, lead and cadmium. It has been shown that fish (i.e. rainbow trout) are exposed to higher cadmium levels and grow at a slower rate than fish exposed to lower levels or none. Moreover, cadmium can potentially alter the productivity and mating behaviours of these fish. Heavy metals can not only affect behaviors, but also the genetic makeup in aquatic organisms. In Canada, a study examined genetic diversity in wild yellow perch along various heavy metal concentration gradients in lakes polluted by mining operations. Researchers wanted to determine as to what effect metal contamination had on evolutionary responses among populations of yellow perch. Along the gradient, genetic diversity over all loci was negatively correlated with liver cadmium contamination. Additionally, there was a negative correlation observed between copper contamination and genetic diversity. Some aquatic species have evolved heavy metal tolerances. In response to high heavy metal concentrations a Dipteran species, *Chironomus riparius*, of the midge family, *Chironomidae*, has evolved to become tolerant to Cadmium toxicity in aquatic environments. Altered life histories, increased Cd excretion, and sustained growth under Cd exposure is evidence that shows that Chironomus riparius exhibits genetically based heavy metal tolerance.

Pesticides

Pesticides are a major source of environmental toxicity. These chemically synthesized agents have been known to persist in the environment long after their administration. The poor biodegradability of pesticides can result in bioaccumulation of chemicals in various organisms along with biomagnification within a food web. Pesticides can be categorized according to the pests they target. Insecticides are used to eliminate agricultural pests that attack various fruits and crops. Herbicides target herbal pests such as weeds and other unwanted plants that reduce crop production.

DDT

Dichlorodiphenyltrichloroethane (DDT) is an organochlorine insecticide that has been banned due to its adverse effects on both humans and wildlife. DDT's insecticidal properties were first discovered in 1939. Following this discovery, DDT was widely used by farmers in order to kill agricultural pests such as the potato beetle, coddling moth and corn earworm. In 1962, the harmful effects of the widespread and uncontrolled use of DDT were detailed by Rachel Carson in her book The Silent Spring. Such large quantities of DDT and its metabolite Dichlorodiphenyldichloroethylene (DDE) that were released into the environment were toxic to both animals and humans.

DDT is not easily biodegradable and thus the chemical accumulates in soil and sediment runoff. Water systems become polluted and marine life such as fish and shellfish accumulate DDT in their tissues. Furthermore, this effect is amplified when animals who consume the fish also consume the chemical, demonstrating biomagnification within the food web. The process of biomagnification has detrimental effects on various bird species because DDT and DDE accumulate in their tissues inducing egg-shell thinning. Rapid declines in bird populations have been seen in Europe and North America as a result.

Humans who consume animals or plants that are contaminated with DDT experience adverse health effects. Various studies have shown that DDT has damaging effects on the liver, nervous system and reproductive system of humans.

By 1972, the United States Environmental Protection Agency (EPA) banned the use of DDT in the United States. Despite the regulation of this pesticide in North America, it is still used in certain areas of the world. Traces of this chemical have been found in noticeable amounts in a tributary of the Yangtze River in China, suggesting the pesticide is still in use in this region.

Sulfuryl Fluoride

Sulfuryl fluoride is an insecticide that is broken down into fluoride and sulfate when released into the environment. Fluoride has been known to negatively affect aquatic wildlife. Elevated levels of fluoride have been proven to impair the feeding efficiency and growth of the common carp (*Cyprinus carpio*). Exposure to fluoride alters ion balance, total protein and lipid levels within these fish, which changes their body composition and disrupts various biochemical processes.

Cyanobacteria and Cyanotoxins

Cyanobacteria, or blue-green algae, are photosynthetic bacteria. They grow in many types of water. Their rapid growth ("bloom") is related to high water temperature as well as eutrophication (resulting from enrichment with minerals and nutrients often due to runoff from the land that induces excessive growth of these algae). Many genera of cyanobacteria produce several toxins. Cyanotoxins can be dermatotoxic, neurotoxic, and hepatotoxic, though death related to their exposure is rare. Cyanotoxins and their non-toxic components can cause allergic reactions, but this is poorly understood. Despite their known toxicities, developing a specific biomarker of exposure has been difficult because of the complex mechanism of action these toxins possess.

Ecotoxicology

Fate, transport and exposure: A pesticide may directly affect something far from the site of application. Pesticides that are bound to soil particles may be carried into streams with runoff. Pesticide drift may travel many miles in the wind. Sunlight, water, microbes, and even air can break down pesticides.

Some pesticides last a long time in the environment, and may pose risks to living things many years after they were last used. Insecticides such as DDT, chlordane, and dieldrin don't break down easily, and they are still found in soil, plants, and animals. Persistent pesticides may travel long distances in the air or water, or even in living organisms such as migrating birds or fish. Researchers have found pesticide residues in alpine lakes and snow, many miles from where the pesticides were applied. Pesticides have even been found in the Arctic and Antarctic environments, probably carried there by currents in the atmosphere or oceans.

Plants may absorb pesticides through their roots or leaves. Animals can be exposed to pesticides directly by breathing them in, getting the pesticides on their skin, or eating them. Pesticides may fool animals; granules may look like food to wild birds especially. Unfortunately, this has actually happened and birds were poisoned as a result. Sometimes an animal's food may be contaminated from pesticide residues in plants or in the tissues of prey. Secondary poisoning can occur if an animal eats another animal that has been fatally poisoned by a pesticide, and predator dies as a result of the poisoned prey. This is also called relay toxicosis.

Some chemicals cross the skin, lungs or gills, and intestine more easily than others. When scientists evaluate the uptake and activity of pesticides in the body, they call it bioavailability. A pesticide's bioavailability depends on whether it is soluble in fat, whether it might be stored in other tissue such as bone or the liver, and how difficult it is for the body to break down the pesticide and excrete it.

Pesticide Build-up in Living Tissue

Pesticide residues build up in organisms and in food webs. Bioaccumulation can occur if residues build up faster than the organism can break them down and excrete them. Bioaccumulation in aquatic animals where the pesticide is taken in from the water is called bioconcentration. If a predator eats many plants and animals that have pesticide residues in their tissues, the predator may suffer from even greater exposure than the prey. Bald eagles, ospreys and peregrine falcons were brought to the brink of extinction because their food sources (fish and other birds) were contaminated with DDE, the breakdown product of the insecticide DDT. The residues built up with each link in the food chain until very high concentrations were present in the eagles, falcons, and ospreys. When residues increase in the food web, the process is called biomagnification.[1] No single exposure for either the prey or the predator is likely to cause injury, but the overall effects can be very harmful.

Effects may be Specific to Time and Place

The timing of exposure can greatly affect how much damage a pesticide might cause. Migrating animals may use a stopover site or staging area only briefly. At that time, those special locations may harbor a large proportion of the population or even the entire species. Other animals form breeding colonies for a few weeks or months of the year. Examples include some species of bats and swifts. If a pesticide is used when and where wildlife are clustered together, much greater harm could result than if that application occurred at another time even in the same place.

Risks may also increase at certain times in the animals' lives. Pesticides may pose greater risks to young animals or animals under stress from migration or breeding. The life stage of a plant may affect its risk of harm. An herbicide may not hurt a seed, or cause only small damage to a large, vigorous plant. However, it might kill a seedling.

Exposure risks may also depend on the conditions in a certain place. For example, barn owls eat voles when they are available. When voles are scarce, barn owls are more likely to eat other rodents such as rats and house mice. Rats and house mice are more likely to carry traces of pesticides. Dead and dying prey may be easier to catch and eat. That means the risks to barn owls depend on what is happening within the rodent community, which affects what prey the barn owls are likely to catch.

Effects on Individuals and Populations of Organisms

Pesticides can affect individual plants and animals in two ways. First, they may cause injury or death after the plant or animal is exposed to the pesticide directly. This might happen if the pesticide drifts onto the plant or animal, the animal breathes in the pesticide, or if the animal drinks or eats something that is contaminated. Plant roots may pick up pesticides in the soil. Any injuries resulting from these exposures are called direct effects. The second way pesticides may cause harm is by changing or killing something the plant or animal needs. For example, pesticides can affect an animal's food supply by killing certain plants or insects. The loss of plant cover may also remove the animal's shelter. Plants could be affected if their pollinators or seed-dispersers are killed. These are indirect effects.

A pesticide does not have to kill an organism to do harm. Instead, a pesticide may have sublethal effects such as making the organism sick, changing its behavior, or changing its ability to reproduce or survive stress. If enough individuals die without leaving behind enough offspring to take their places, the population gets smaller. For example, young salmon exposed to pesticides do not grow or survive as well as unexposed fish. Over time, this could affect salmon population numbers. Pesticides could affect a population through direct or indirect, as well as lethal or sublethal effects.

Community Effects

Effects can also occur on larger ecological scales than that of the individual. For example, predator-prey relationships may be changed by pesticides and other contaminants.[11] If the predatory wasps are more affected by an insecticide than the pests they feed on, the pest population may grow. The population of pests will often recover faster than the populations of predators following pesticide applications in agriculture.

When pesticides remove one of the species at the bottom of a food web, many other species may be affected. In this example of community-level effects, spraying for mosquitoes with Bacillus thuringiensis israelensis (Bti) reduced the populations of midges and mosquitoes, the favorite food of house martins. House martins in treated areas made fewer trips back to their nest with food, and raised fewer young, than house martins living in untreated areas. Spiders and dragonflies declined in treated areas, probably because they also eat midges and mosquitos.

Declines in the number of one species may also affect plants or other organisms. For example, if a butterfly's host plant is affected by pesticides, they may not have enough places to lay eggs. If a pollinator species is lost, plants may not be able to set enough seed to maintain their numbers. These indirect effects can be very difficult to predict without doing experiments.

Researchers put together large containers for stream insects and earthworms, and they added leaves from trees that were treated with the insecticide imidacloprid. Aquatic insects and microbes decomposed fewer leaves from the treated trees, and the earthworms lost weight compared to controls. Treated leaves were therefore not as quickly broken down by earthworms as the control leaves were. This study demonstrated a sublethal effect on decomposers by affecting their feeding behavior, which led to indirect effects on the whole community because of the slower breakdown of the leaves.

In another study, researchers created ponds made from watering tanks, placing plants and animals inside that would be found in an ordinary pond. When they sprayed some of the tanks with the insecticide malathion, many effects occurred. The number of tiny aquatic animals called zooplankton declined in the treated tanks. Zooplankton feed on phytoplankton, tiny floating plants. Phytoplankton increased when zooplankton densities declined, and blocked light penetration to the bottom of the tanks. Algae and other organisms growing on the bottom died from lack of light. Leopard frog tadpoles had less food, and grew more slowly. This made them more likely to die as the ponds dried up. An effect that ripples through a community like this is called a trophic cascade.

Another study studied the relationship between parasitic flatworm infections in leopard frogs and pesticides in water. The concentration of atrazine, a common herbicide, was directly related to the number of parasites infecting the leopard frogs. When atrazine was added to the tanks, it killed the phytoplankton. More sunlight reached the bottoms of the tanks, allowing periphyton to grow. More snails were able to live in those tanks. The flatworms use snails as hosts before they infect the frogs, so more snails meant more flatworms. The scientists concluded that the atrazine indirectly increased parasitism in the frogs by increasing the population of snail hosts.

At the Ecosystem Level

Pesticides and contaminants may affect more than just the populations of animals and plants that make up a community. They may also affect basic processes like nutrient cycling or the formation of soil. For example, nitrogen cycling may be affected if pesticides impact the bacteria and fungal communities in soil. There may be a time lag between the pesticide exposure and the ultimate effects. The pesticide could be gone before the damage it caused also disappears.

Ecotoxicity Testing

- Acute and chronic toxicity tests are performed for terrestrial organisms including avian, mammalian, nontarget arthropods, and earthworms.

- The Organization for Economic Cooperation and Development (OECD) test guideline has developed specific tests to test toxicity level in organisms. Ecotoxicological studies are generally performed in compliance with international guidelines, including EPA, OECD, EPPO, OPPTTS, SETAC, IOBC, and JMAFF.

- LC_{50} is the acute toxicity test that tests for the concentrate of tissue at which it is lethal to 50% within the test-specified time. The test may start with eggs, embryos, or juveniles and last from 7 to 200 days.

- EC_{50} is the concentration that causes adverse effects in 50% of the test organisms (for a binary yes/no effect such as mortality or a specified sublethal effect) or causes a 50% (usually) reduction in a non-binary parameter such as growth.

- Endocrine Disruptor Screening Program (EDSP).

- Tier 1 screening battery.

- Endangered species assessments.

- Persistent, Bioaccumulative, and Inherently Toxic (PBiT) assessments using the Quantitative Structure-Activity Relationships (QSARs) to categorize regulated substances.

- Bioaccumulation in fish using the Bioconcentration Factor (BCF) methods.

Classification of Ecotoxicity

Total amount of acute toxicity is directly related to the classification of toxicity.

< 1 part per million → Class I

1–10 parts per million → Class II

10–100 parts per million → Class III

Aquatic Toxicology

Aquatic toxicology is the study of the adverse effects of toxins and their activities on aquatic ecosystems. Aquatic toxicologists assess the condition of aquatic systems, monitor trends in conditions over time, diagnose the cause of damaged systems, guide efforts to correct damage, and predict the consequences of proposed human actions so the ecological consequences of those actions can be considered before damage occurs. Aquatic toxicologists study adverse effects at different spatial, temporal, and organizational scales. Because aquatic systems contain thousands of species, each of these species can respond to toxicants in many ways, and interactions between these species can be affected.

Historically, this discipline has used toxicity tests to identify the harmful effects. Standard tests evaluate dose-response relationships (toxicity at different concentrations) and mechanisms of action in a variety of organisms that are representative of different ecosystem niches. These tests may evaluate the response of individuals or populations to varying concentrations of the chemical. The dose-response relationship is based on the following three assumptions:

1. The response (toxicity) is due to the chemical administered.

2. The magnitude of the response (toxicity) is related to the dose.

3. There exists both a quantifiable method for measuring and a precise means of expressing toxicity.

Types of Effects

Effects may be of such minor significance that the organism can function normally. However, under stressful conditions (i.e., pH change, low dissolved oxygen, high temperatures, changes in hardness, etc.), the same chemical exposure may become very lethal. The toxicity of some chemicals may also be enhanced or mitigated in the presence of other chemicals. In addition to killing the organisms, some pesticides can have negative but non-lethal effects on individual organisms and populations, such as reduced reproduction, reduced mobility to escape predation, or alterations in behavior.

Toxicity Measurement and Estimation

One common measurement used to describe toxicity of pesticides to organisms is the LC_{50}, or the statistically derived concentration in water that can be expected to cause death in 50% of the animals exposed.

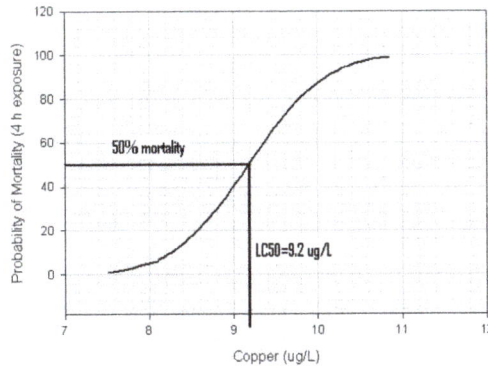

Figure: Sample toxicity curve showing water flea mortality. The sigmoidal line represents actual percent mortality. The $LC_{50} = 9.2$ μg/L in this example.

For estimation of non-lethal effects on processes such as growth and reproduction, the EC_{50}, or the statistically derived concentration in water that can be expected to cause a reduction of 50% in the process being measured, is used.

Aquatic Toxicity Tests

Aquatic toxicology tests (assays): toxicity tests are used to provide qualitative and quantitative data on adverse (deleterious) effects on aquatic organisms from a toxicant. Toxicity tests can be used to assess the potential for damage to an aquatic environment and provide a database that can be used to assess the risk associated within a situation for a specific toxicant. Aquatic toxicology tests can be performed in the field or in the laboratory. Field experiments generally refer to multiple species exposure and laboratory experiments generally refer to single species exposure. A dose response relationship is most commonly used with a sigmoidal curve to quantify the toxic effects at a selected end-point or criteria for effect (i.e. death or other adverse effect to the organism). Concentration is on the x-axis and percent inhibition or response is on the y-axis.

The criteria for effects, or endpoints tested for, can include lethal and sublethal effects.

There are different types of toxicity tests that can be performed on various test species. Different species differ in their susceptibility to chemicals, most likely due to differences in accessibility, metabolic rate, excretion rate, genetic factors, dieteary factors, age, sex, health and stress level of the organism. Common standard test species are the fathead minnow (Pimephales promelas), daphnids (*Daphnia magna, D. pulex, D. pulicaria, Ceriodaphnia dubia*), midge (Chironomus tentans, C. ruparius), rainbow trout (Oncorhynchus mykiss), sheepshead minnow (Cyprinodon variegatu), zebra fish (*Danio rerio*), mysids (Mysidopsis), oyster (Crassotreas), scud (Hyalalla Azteca), grass shrimp (Palaemonetes pugio) and mussels (*Mytilus galloprovincialis*). As defined by ASTM, these species are routinely selected on the basis of availability, commercial, recreational, and ecological importance, past successful use, and regulatory use.

A variety of acceptable standardized test methods have been published. Some of the more widely accepted agencies to publish methods are: the American Public Health Association, U.S. Environmental Protection Agency, American Society for Testing and Materials, International Organization for Standardization, Environment Canada, and Organization for Economic Cooperation and Development. Standardized tests offer the ability to compare results between laboratories.

There are many kinds of toxicity tests widely accepted in the scientific literature and regulatory agencies. The type of test used depends on many factors: Specific regulatory agency conducting the test, resources available, physical and chemical characteristics of the environment, type of toxicant, test species available, laboratory vs. field testing, end-point selection, and time and resources available to conduct the assays are some of the most common influencing factors on test design.

Exposure Systems

Exposure systems are four general techniques the controls and test organisms are exposed to the dealing with treated and diluted water or the test solutions:

- Static: A static test exposes the organism in still water. The toxicant is added to the water in order to obtain the correct concentrations to be tested. The control and test organisms are placed in the test solutions and the water is not changed for the entirety of the test.

- Recirculation: A recirculation test exposes the organism to the toxicant in a similar manner as the static test, except that the test solutions are pumped through an apparatus (i.e. filter) to maintain water quality, but not reduce the concentration of the toxicant in the water. The water is circulated through the test chamber continuously, similar to an aerated fish tank. This type of test is expensive and it is unclear whether or not the filter or aerator has an effect on the toxicant.

- Renewal: A renewal test also exposes the organism to the toxicant in a similar manner as the static test because it is in still water. However, in a renewal test the test solution is renewed periodically (constant intervals) by transferring the organism to a fresh test chamber with the same concentration of toxicant.

- Flow-through: A flow-through test exposes the organism to the toxicant with a flow into the test chambers and then out of the test chambers. The once-through flow can either be intermittent or continuous. A stock solution of the correct concentrations of contaminant must be previously prepared. Metering pumps or diluters will control the flow and the volume of the test solution, and the proper proportions of water and contaminant will be mixed.

Types of Tests

Acute tests are short-term exposure tests (hours or days) and generally use lethality as an endpoint. In acute exposures, organisms come into contact with higher doses of the toxicant in a single event or in multiple events over a short period of time and usually produce immediate effects, depending on absorption time of the toxicant. These tests are generally conducted on organisms during a specific time period of the organism's life cycle, and are considered partial life cycle tests. Acute tests are not valid if mortality in the control sample is greater than 10%. Results are reported in EC50, or concentration that will affect fifty percent of the sample size.

Chronic tests are long-term tests (weeks, months years), relative to the test organism's life span (>10% of life span), and generally use sub-lethal endpoints. In chronic exposures, organisms come into contact with low, continuous doses of a toxicant. Chronic exposures may induce effects to acute exposure, but can also result in effects that develop slowly. Chronic tests are generally considered full life cycle tests and cover an entire generation time or reproductive life cycle ("egg to egg"). Chronic tests are not considered valid if mortality in the control sample is greater than 20%. These results are generally reported in NOECs (No observed effects level) and LOECs (Lowest observed effects level).

Early life stage tests are considered as subchronic exposures that are less than a complete reproductive life cycle and include exposure during early, sensitive life stages of an organism. These exposures are also called critical life stage, embryo-larval, or egg-fry tests. Early life stage tests are not considered valid if mortality in the control sample is greater than 30%.

Short-term sublethal tests are used to evaluate the toxicity of effluents to aquatic organisms. These methods are developed by the EPA, and only focus on the most sensitive life stages. Endpoints for these tests include changes in growth, reproduction and survival. NOECs, LOECs and EC50s are reported in these tests.

Bioaccumulation tests are toxicity tests that can be used for hydrophobic chemicals that may accumulate in the fatty tissue of aquatic organisms. Toxicants with low solubilities in water generally can be stored in the fatty tissue due to the high lipid content in this tissue. The storage of these toxicants within the organism may lead to cumulative toxicity. Bioaccumulation tests use bioconcentration factors (BCF) to predict concentrations of hydrophobic contaminants in organisms. The BCF is the ratio of the average concentration of test chemical accumulated in the tissue of the test organism (under steady state conditions) to the average measured concentration in the water.

Freshwater tests and saltwater tests have different standard methods, especially as set by the regulatory agencies. However, these tests generally include a control (negative and positive), a geometric dilution series or other appropriate logarithmic dilution series, test chambers and equal numbers of replicates, and a test organism. Exact exposure time and test duration will depend on type of test (acute vs. chronic) and organism type. Temperature, water quality parameters and light will depend on regulator requirements and organism type.

In the US, many wastewater dischargers (e.g., factories, power plants, refineries, mines, municipal sewage treatment plants) are required to conduct periodic whole effluent toxicity (WET) tests under the National Pollutant Discharge Elimination System (NPDES) permit program, pursuant to the Clean Water Act. For facilities discharging to freshwater, effluent is used to perform static-acute multi-concentration toxicity tests with *Ceriodaphnia dubia* (water flea) and *Pimephales promelas* (fathead minnow), among other species. The test organisms are exposed for 48 hours under static conditions with five concentrations of the effluent. The major deviation in the short-term chronic effluent toxicity tests and the acute effluent toxicity tests is that the short-term chronic test lasts for seven days and the acute test lasts for 48 hours. For discharges to marine and estuarine waters, the test species used are sheepshead minnow (*Cyprinodon variegatus*), inland silverside (*Menidia beryllina*), *Americamysis bahia*, and purple sea urchin (*Strongylocentrotus purpuratus*).

Sediment Tests

At some point most chemicals originating from both anthropogenic and natural sources accumulate in sediment. For this reason, sediment toxicity can play a major role in the adverse biological effects seen in aquatic organisms, especially those inhabiting benthic habitats. A recommended approach for sediment testing is to apply the Sediment Quality Triad (SQT) which involves simultaneously examining sediment chemistry, toxicity, and field alterations so that more complete information can be gathered. Collection, handling, and storage of sediment can have an effect on bioavailability and for this reason standard methods have been developed to suit this purpose.

Occupational Toxicology

The work environment with its chemical and biologic hazards plays a role in the occurrence of adverse human health effects. Occupational toxicology is the application of the principles and methodology of toxicology toward chemical and biologic hazards encountered at work. The objective of the occupational toxicologist is to prevent adverse health effects in workers that result from their work environment. Because the work environment often presents exposures to complex mixtures, the occupational toxicologist must also recognize exposure combinations that are particularly hazardous.

Many of the chemicals that occupational toxicologists must deal don't cause an appreciable risk to health when they are present at low levels in food, consumer products and the environment. However, workers may be exposed to these chemicals at considerably higher levels than the general public, so the consequences of human exposure are potentially the most serious in the work place. Occupational toxicology is not only important in chemical factories, but is just as relevant in high-street bakeries handling flour dust, or hairdressers using hair dyes. They also advise government on legal controls necessary to ensure that chemicals are handled and used safely.

Occupational toxicologists must understand the potential toxicity (hazard) posed by a particular substance. They must then assess the risks to human health in specific occupational settings, taking into account the level duration and route of exposure and any other factors that influence the way that workers handle the substance.

Having prepared a risk assessment, occupational toxicologists can advise on suitable working conditions and equipment needed to protect the workers exposed to the substance. They can also offer advice on how the substance should be safely handled and stored. It may be necessary to work with occupational hygienists and physicians to draw up rules that govern the use of chemicals, and to communicate this advice to management and employees.

In the unfortunate event of an accident involving a chemical at work, occupational toxicologists may be asked to advice on how to treat any workers that have been accidentally exposed to the chemical, or what action should be taken if a chemical is accidentally released into the environment.

Workplaces are complex environments where many chemicals may be in use at the same time, so it is important that occupational toxicologists understand and can recognise the potential dangers of simultaneous exposure to more than one type of chemical.

First, the clinical expressions of occupationally induced diseases are often indistinguishable from those arising from nonoccupational causes. Second, there may be a long interval between exposure and the expression of disease. Third, diseases of occupational origin may be multifactorial with personal or other environmental factors contributing to the disease process.

Determinants of Dose

Dose is defined as the amount of toxicant that reaches the target tissue over a defined time span. In occupational environments, exposure is often used as a surrogate for dose. The response to a toxic agent is dependent on both host factors and dose. Figure illustrates the pathway from exposure to subclinical disease or adverse health effect and suggests that there are important modifying factors: contemporaneous exposures, genetic susceptibility, age, gender, nutritional status, and behavioral factors. These modifying factors can influence whether a worker remains healthy, develops subclinical disease that is repaired, or progresses to illness. As illustrated in Figure, the dose is a function of exposure concentration, exposure duration, and exposure frequency. Individual and environmental characteristics also can affect dose.

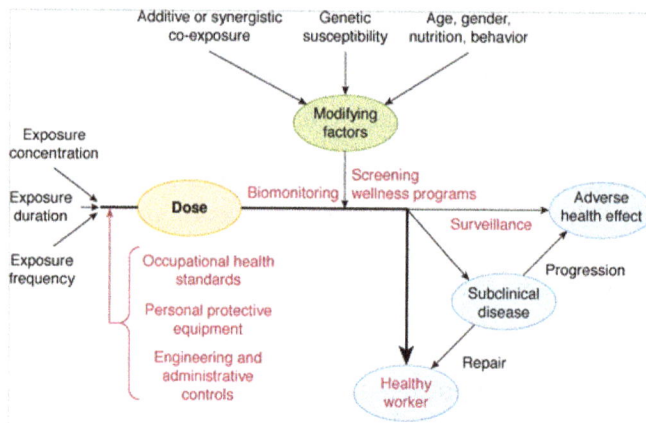

Medical Toxicology

Medical toxicology is a branch of toxicology that focuses on the diagnosis, management and prevention of adverse health effects resulting from medications or from exposure to toxic chemical or biological agents. Medical toxicology focuses on the diagnosis, management and prevention of poisoning due to drugs, occupational and environmental toxins, and biological agents. Examples of exposures commonly evaluated by medical toxicologists include acute drug overdoses, envenomations, ingestions of food borne or plant and mushroom toxins, hazardous exposures to chemical products, and the management of drug withdrawal syndromes. Medical toxicologists practice in a variety of professional settings including the direct treatment and consultation of acutely poisoned patients in emergency departments or intensive care units, poison control center management, industry and commerce, as well as government regulatory bodies.

Some Examples of Problems Evaluated by Medical Toxicologists

Medical toxicologists are involved in the care of people who come into contact with drugs, substances or other agents causing potentially adverse health effects. This entails expertise in many areas, such as:

- Unintentional and intentional overdoses of such agents as:
  - Therapeutic drugs including antidepressants, cardiac medications and many others;
  - Over-the-counter medicines;
  - Drugs of abuse;
- Exposure to industrial chemical products and environmental hazards such as:
  - Pesticides;
  - Heavy metals;
  - Household products;
  - Toxic gases;
  - Toxic alcohols;
  - Other industrial and environmental agents, including radiation exposures;
- Drug abuse management including:
  - Inpatient care for acute withdrawal states from addictive agents such as alcohol and drugs of abuse;
  - Outpatient addiction medicine treatment;
- Diagnosis and management of exposures such as:
  - Snake, scorpion and spider envenomations;
  - Marine toxins;
  - Ingestion of food-borne toxins;
  - Ingestion of toxic plants;
- Independent medical examinations, assessing injury or disability resulting from toxic exposures.

Medical Toxicologists work in a variety of settings including:

- Emergency departments and in-patient units where they directly treat acutely poisoned patients;

- Outpatient clinics and occupational health settings where they evaluate the health impact from exposure to toxic substances in the home or workplace.

- National and regional poison control centers where they provide medical direction for health professionals, personal responders and the general public.

- Academic institutions where they are involved in teaching, research, and improving evidence-based patient care.

- Industry and commerce where they contribute to pharmaceutical research and development, product safety, occupational health services, and regulatory compliance.

- Governmental agencies where they provide toxicology expertise at all levels from local health departments to federal entities.

- Clinical and forensic laboratories where they aid in the design, conduction and interpretation of diagnostic tests and forensic studies.

Pharmacotoxicology

Pharmacotoxicology is the science that involves the study of the prevention and control of adverse health effects caused due to toxic exposure to pharmaceutical products. It is separated into the two principal categories viz. pharmacokinetics and pharmacodynamics. The exhibition of toxicity is often a result of the bioactivation of drugs, drug-drug interaction or a human immune response to a pharmaceutical agent. It may also happen that the drugs have a toxic response owing to its interaction with alternate targets. The field also focuses on the prevention and treatment of such pharmaceutically induced side-effects.

Forensic Toxicology

Forensic toxicology is a part of the science of pharmacology, which is concerned with the quantities and effects of various drugs and poisons on human beings. In forensic toxicology the main interest is the extent to which drugs and poisons may have contributed to impairment or death. More than half of the cases received by forensic toxicologists involve drinking alcohol and driving. Every state and the Federal Government has laws that prohibit drinking and driving and set levels above which a person is either impaired or OUIL. Forensic toxicologists are called upon to determine the level of alcohol present in the body and sometimes, the level at a previous time and the effects on the person. In cases involving drugs and poisons, forensic toxicologists usually only get involved when death has occurred. The toxicologist works with the medical examiner or coroner to help determine the cause and manner of death. The toxicologist will use data about what drugs are present and at what levels at the time of death, along with drug usage history and general health, to determine the role that drugs or poisons played in death.

Forensic toxicology continues to be a dynamic field with evolving technology applications. Tandem mass spectrometry methods, particularly liquid chromatography-tandem mass spectrometry (LC-MS/MS), have grown in importance. These technologies provide greater sensitivities and flexibility for the detection of larger and more polar compounds that are difficult or impossible to analyze with gas chromatographic methods. Continued development of immunoassays for a wider range of compounds has kept immunoassays a vital part of the analytical armamentarium. Time of flight mass spectrometry (TOF-MS) has also become more important in the field. The advent of more, larger peptide-based drugs will present analytical challenges for forensic toxicologists. Also the rising field of pharmacogenomics and the concept of the genetic autopsy may alter dramatically the interpretation of drug concentrations. As more evidence is gained about the interplay between an individual's geno- and phenotypes and their mxetabolic capacity, changes in dosing and in the interpretation of what is viewed to be toxic and therapeutic are likely to occur.

A careful development and selection of new appropriate methods and instrumentation as well as their suitable application is one major crucial point regarding the claims and laboratory performance in forensic toxicology. Two kinds of approaches, methods, and corresponding equipment need to be further developed, validated, if possible, and maintained regarding open tasks to be carried out:

1. Approved general unknown analysis and selective screening methods.

2. Various quantitative and confirmation methods need to be applied according to specific requests and in cases in which specific qualitative results were previously obtained by general unknown analysis.

There are a practically unlimited number of poisons that may be present in individual cases and under particular circumstances. Therefore, forensic toxicology is a scientific discipline in which permanent efforts to complete and improve the methods of poison detection show its close relation to raising quality.

Forensic toxicology can even identify poisons and hazardous chemicals. The chemical makeup of each substance is studied and they are also identified from different sources such as urine or hair. Forensic toxicology deals with the way that substances are absorbed, distributed or eliminated in the body – the metabolism of substances. When learning about drugs and how they act in the body, forensic toxicology will study where the drug affects the body and how this occurs.

Obtaining Samples for Toxicology Testing

Before toxicology testing can go forward, samples need to be taken. You might be surprised to know just how many parts of your body can produce samples that are effective for identifying drugs. One example is urine, which is commonly used in forensic toxicology. It's an easy sample to obtain and relatively rapid and non-invasive. It can show substances even several weeks after their ingestion. One example would be the drug marijuana, which can be detected even two weeks following use of the drug. When a urine sample is taken, however, there are sometimes rules and regulations around how the sample is collected. If the testing was related to workplace drug testing, a person could substitute a sample from someone else that would then show a negative result. For this reason, there are sometimes parameters around reasonable supervision when a person has to provide a urine sample.

Blood samples are another body sample used for forensic toxicology. A huge range of toxic substances can be tested from a blood sample. You may already be familiar with blood alcohol testing used to assess if a person was driving under the influence of alcohol. This type of testing is important in assessing if a driver is above the legal limit and it is also used to prove a case in court.

Hair samples are a good way to test for substance abuse that has occurred over the long-term. After a person ingests a chemical, it ends up in the hair, where it can provide forensic toxicologists with an estimate of the intensity and duration of drug use. Hair testing is even offered quite widely by companies that allow you to mail in a hair sample and check off the drugs you want checked. Saliva is another way that forensic toxicologists can test for drugs. It does, however, depend on the drug in terms of identifying its concentration. One of the more unusual sounding but interesting ways that the human body can be used for forensic toxicology involves the gastric contents in a deceased person. During the autopsy, a sample of the person's gastric contents can be analysed, which then allows the forensic toxicologist to assess if the person took any pills or liquids before their death. The brain, liver and spleen can even be used during toxicology testing.

Forensic Toxicology Applications

While there are many uses for forensic toxicology testing, the most familiar one to most people is likely to be drug and alcohol testing. This type of testing is commonly performed in the transportation industry and in workplaces. Another use is for drug overdoses, whether these are intended or accidental. People who drive with a blood alcohol concentration over the accepted legal limit can also be assessed through toxicology testing. Another application of forensic toxicology relates to sexual assault that involves the use of drugs. Various drugs are used today for the purposes of rendering the victim unable to fight the attacker, who then proceeds to sexually assault the victim. Through toxicology testing, a victim can find out what drug was given and can then be treated accordingly.

There are a lot of substances and poisons in our world – many of which impact how we function in work and society. For some people, these substances can influence their death. Fortunately, forensic toxicology testing allows forensic scientists to identify substances and determine a pattern of use. In this way, a forensic toxicologist can provide closure on the 'what if' of a person's drug habits or perhaps some mystery surrounding their death.

Entomotoxicology

Entomotoxicology, a relatively new branch of forensic entomology, deals with the use of insects in detecting drugs and other toxins in decomposing tissues. This also examines the application of analytical techniques to carrion feeding insects in order to identify drugs and toxins present in any intoxicated tissues. These insects, as they feed on corpses, ingest, incorporate, and accumulate drugs and metabolites in the cadaver into their own body. The drugs and toxins are locked in the cuticle of the larvae or empty pupal cases; therefore, they are useful sources for toxicological analysis. Their use as an alternative matrix for drug detection is well documented and recommended when conventional matrices such as blood, urine, and tissues are not available.

Figure: Calliphora vomitoria (adult fly)

The pharmacokinetics of drugs in insects depends on species, developmental stage, and feeding activity. Apart from necrophagous species, bioaccumulations have also been observed in parasitoids, predators, and omnivorous species. In entomotoxicological investigations, necrophagous species belonging to Coleoptera (beetles) and Diptera (flies) are highly recommended as they are the first to colonize a corpse and these species are commonly encountered in crime scenes. Calliphoridae, also known as the blowflies (order: Diptera) are predominant and forensically important insects as they are generally the first arthropods to locate and oviposit on a corpse appearing within minutes following death.[2] Blowflies detect carcases primarily through the odor of decaying tissues. Other insects include sarcophagidae (flesh flies, order: Diptera) which generally arrive after the blowflies and muscidae (houseflies, order: Diptera). Therefore, the knowledge of local insect assemblages and their growth rate as well as population dynamics is important for application of entomotoxicology for forensic purposes.

A drug or toxin can be detected in larvae when its rate of absorption exceeds the rate of elimination. However, it is not clearly known how larvae bioaccumulate or eliminate drugs, and how these affect larval development. Early studies during the 1970s have focused on the detection of metals and in the 80s studies focused on pesticides and drugs in insects. An example was the analysis of maggots recovered from thoracic and abdominal cavities of a 22 year old female corpse which was found in a bad state of decomposition. Phenobarbital was detected, and since then larvae found on cadavers have been used for identification of drugs present in a corpse.

The type of drug present and their concentration could affect the growth rate of fly species, therefore understanding the effects of drugs and toxins on their development are essential. The determination of time since death from entomological evidence is most valuable in cases of badly decomposed or skeletonized bodies. By taking into consideration the insect succession during the decomposition and the stage of development (egg, larva, pupa, or adult) at which insect was found, it is possible to calculate the minimum time since death. Previous studies have shown that the presence of drugs could accelerate, retard, or have no influence on the development rates of different larvae. Some of these effects depend on the concentration (e.g. methamphetamine and cocaine) of the drug while others simply depend on its presence.

The drug metabolism, absorption, and elimination as well as possible drug accumulation and localization in insects are still not fully understood. There are also variations in drug quantities reported in the different larval and pupal stages of insects which pose difficulty in quantification. The presence of metabolites could add further complexity in the interpretation of quantitative results as metabolites could have been from the corpse and due to the metabolism in the insect itself. It is therefore clear that the metabolism of drugs and excretion in the different developmental stages of insects should be explored in more detail to get an idea of how and to what extent drugs are incorporated in insect tissues.

In terms of sample preparation and analysis, samples are homogenized and subjected to extraction techniques developed for the analysis of specimens such as hair and nails. Both solid phase and liquid—liquid extractions have been commonly used for sample preparation. Various enzyme immunoassays, chromatographic and mass spectrometric techniques have been used for the detection of drugs. This is similar to conventional toxicology sample analysis such as blood and urine.

Larvae, particularly those that are actively feeding on human bodies, provide a potentially valuable source of information in forensic investigations. Due to the increased drug related deaths over the last decade, entomotoxicology is gaining interest within the medico-legal field. The development of advanced analytical techniques has given this field a great boost. These alternative samples such as maggots and larvae make excellent qualitative toxicological specimens even though there have been issues with quantification.

On the other hand, it is very difficult to obtain real case reports and databases for further research in entomotoxicology.[4] Post mortem drug distribution and drug stability in humans and insects are not always well known, and this complicates interpretation. The use of human tissues such as liver to rear larvae experimentally is also ethically questionable. Therefore, in the last few years, in vitro research has been conducted trying to find a relationship between drug concentrations in substrates and insects reared on that substrate, and to increase the knowledge of insect development. On account of the above remarks, further entomotoxicological research should be carried out, especially focusing on bioaccumulation and insect metabolism of drugs.

Effects of Toxins on Arthropods

Drugs can have a variety of effects on development rates of arthropods. Morphine, heroin, cocaine, and methamphetamine are commonly involved in cases where forensic entomology is used. The stages of growth for insects provides a basis for determining a cause in altered cycles in a specific species. An altered stage in development can often indicate toxins in the carrion on which the insects are feeding. Beetles (Order: Coleoptera) and beetle feces are often used in entomotoxicology, but the presence of toxins is often the result of the beetles' feeding on fly larvae that have been feeding on the carrion containing toxic substances. Flies (Order: Diptera) are the most commonly used insect in entomotoxicology.

Through the study of *Sarcophaga (Curranea) tibialis* larvae, barbiturates were found to increase the length of the larval stage of the fly, which will ultimately cause an increase in the time it takes to reach the stage of pupation. Morphine and heroin were both believed to slow down the rate of fly development. However, closer examination of the effects of heroin on fly development has shown that it actually speeds up larval growth and then decreases the development rate of the pupal stage. This actually increases the overall timing of development from egg to adult. Research of *Lucilia sericata* (Diptera: Calliphoridae), reared on various concentrations of morphine injected meat, found higher concentrations of morphine in shed pupal casings than in adults. Cocaine and methamphetamine also accelerate the rate of fly development.

Some effects depend on the concentration of the toxin while others simply depend on its presence. For example, cocaine (at the lethal dose) causes larvae to "develop more rapidly 36 (to 76) hours after hatching". The amount of growth depends on the concentration of cocaine in the area being fed upon. The amount of methamphetamine, on the other hand, affects the rate of pupal development. A lethal dose of methamphetamine increases larval development through approximately the first two days and afterwards the rate drops if exposure remains at the median lethal dosage. The presence of methamphetamine was also found to cause a decrease in the maximum length of the larvae.

Along with changes in development rates, extended periods of insect feeding refrain and variation in the size of the insect during any stage of development, can also indicate the presence of toxic substances in the insect's food source.

Examples of use

Since J.C. Beyer and his partners first demonstrated the ability of toxins to be recovered from maggots feeding on human remains in 1980, the use of entomotoxicology in investigations has made an emergence into the field of forensic entomology. An example of one such case involved the discovery of a 22-year-old female with a history of suicide attempts found 14 days after her death. Due to the body's advanced stage of decomposition, no organ or tissue samples were viable to screen for toxins. Through gas chromatography (GC) and thin-layer chromatography (TLC) analysis of *Cochliomyia macellaria* (Diptera: Calliphoridae) larvae found feeding on the woman's body, phenobarbital was detected and perceived to have been in the woman's system upon death.

Drug Abuse Detected

In France, Pascal Kintz and his colleagues were able to demonstrate the use of entomotoxicology to detect toxins that were not discovered during the analysis of body tissues and fluids of a body found roughly two months after death. A liquid chromatography analysis on organ tissue and Calliphoridae larvae found at the scene revealed the existence of five prescription medications. Triazolam, however, was only detected in the analysis of maggots and not in organ tissue samples. Comparative research showed increased sensitivity of toxicological analysis of Diptera samples over decomposed body tissues. A similar case involved the discovery of the remains of a 29-year-old known to abuse drugs, last seen alive five months prior. Through the use of GC and GC-MS techniques, Nolte and his partners discovered the presence of cocaine in decomposed muscle tissue and in maggots found on the body. However, due to the severity of decomposition of the muscle tissue, more suitable drug samples (devoid of decomposition byproducts) were reared from the maggots.

Aid Determination of Origin

Pekka Nuorteva presented the case of a young woman found severely decomposed in Inkoo, Finland. Diptera larvae recovered from the body were reared to adulthood and found to contain low levels of mercury, indicating that the woman came from an area of comparatively low mercury pollution. This assumption was proven correct once the woman was identified and found to have been a student in Turku, Finland. This case demonstrated the ability of toxicological analysis to help determine origin. This case applied Nuorteva's research involving mercury and its effect on maggots. Through experimentation, it was determined that maggots (fed on fish containing mercury) possessed levels of mercury in their tissue of even greater concentration than in the tissue of the fish. Nuorteva also discovered that the presence of mercury in the maggots systems hindered their ability to enter into the pupal stage.

Toxin Confounding of Postmortem Interval Estimate

Through the analysis of specific cases, it was revealed that toxins present in a person's body upon death can confound postmortem interval estimations. An example of such a case, reported by Gunatilake and Goff, concerned the discovery of a 58-year-old male with a history of attempted suicides found dead in a crawl space in Honolulu, Hawaii last seen eight days prior. Two species of Diptera (Calliphoridae), *Chrysomya megacephala* and *Chrysomya rufifacies*, found on the corpse and tissue samples from the body revealed malathion. Investigators found it abnormal that, given the conditions, there were only two fly species found on the body and that these species revealed a postmortem interval of five days. Thus it was determined that the presence of the organophosphate malathion in the man's system delayed oviposition for a few days.

Paul Catts analyzed a case in Spokane, Washington where maggots rendered differing postmortem estimations. A 20-year-old female victim was found stabbed to death and laying in an open environment surrounded by trees. Most of the oldest maggots found on the body were approximately 6–7 mm long which suggested that they were roughly seven days old. There was, however, a very strange exception which was the retrieval of a 17.7 mm maggot which suggested an age of 3 weeks. After ruling out the possibility that the maggot had traveled onto the corpse from carrion nearby, it was assumed that there was no conceivable way a 3 week old maggot could have been present on the corpse. Later investigations revealed that the woman had snorted cocaine shortly before her death and that the 17.7 mm maggot must have feed in the woman's nasal cavity. Research revealed that maggot development can be sped up by the ingestion of cocaine.

Use of Shed Casings and Insect Feces

Not only are tissues from maggots used to detect toxins, shed casings and insect feces have also been used to detect and identify toxins present in corpses upon death. An instance of this finding was demonstrated by Edward McDonough, a medical examiner in Connecticut. A mummified corpse of a middle-aged woman was found inside of her home. Prescription medicine bottles were found with labels identifying the following drugs: ampicillin, Ceclor, doxycyline, erythromycin, Elavil, Lomotil, pentazocine, and Tylenol 3. McDonough performed toxicological analyses on stomach contents and dried sections of brain and found lethal levels of amitriptyline and nortriptyline. Insect feces, shed pupal cases of *Megaselia scalaris* (Diptera: Phoridae), and shed larval skins of *Dermestes maculates* (Coleoptera: Dermestidae) were gathered from the corpse at the scene. McDonough sent these to an FBI lab which broke down the complex structures of the samples using strong acids and bases and freed the toxins for analysis. The cast pupal cases and larval skins were also found to contain amitriptyline and nortriptyline. Larger concentrations were discovered in the pupal cases because phorid flies prefer to feed on softer tissues. The hide beetle larval skins revealed lower concentrations of the drugs because these beetles prefer to feed on dry, mummified bodies. The use of pupal cases and larval skins allows for investigators to detect toxins in a body years after death.

Limitations

Further research should be conducted in order to fill the gaps in entomotoxicology. Such areas as bioaccumulation, insect metabolism of drugs, and quantitative analyses of insect evidence have only begun to be researched. Because it is a relatively new branch of forensic entomology, entomotoxicology has its limitations. According to Pounder's research, there is no correlation between the drug concentration in tissue and the larvae feeding on that tissue. Entomological specimens make for excellent qualitative toxicological specimens. There is, however, a lack of research in the way of developing an assessment that can quantify the concentration of a drug in tissue using entomological evidence. One reason for this is that a drug can only be detected in larvae when the rate of absorption exceeds the rate of elimination. demonstrated this theory using *Calliphora vicina* larvae reared on human skeletal muscle obtained from cases of co-proxamol and amitriptyline overdose. Samples of pupae and third instar larvae no longer contained concentrations of the drugs, suggesting that drugs do not bioaccumulate over the entire life-cycle of larvae. This leads entomologists to theorize that toxins are eliminated from the larvae's system over time if they are not receiving a constant supply of the toxin.

Nanotoxicology

Nanotoxicology is the study of the toxicity of nanomaterials. It is a branch of bionanoscience which deals with the study and application of toxicity of nanomaterials. Nanotoxicology is a sub-specialty of particle toxicology. It was proposed to address the adverse effects likely to be caused by nanomaterials that seem to possess toxicity effects that are unusual and not seen with larger particles. Nanotechnology is a double-edge sword, the same novel properties making nanoparticles attractive, makes them potentially toxic. Given the excitement associated with all of the nanotechnology applications, evaluating the potential hazards related to exposures to nanoscale materials and its products has become an emerging area in toxicology and health risk assessment. Typical nanoparticles that have been studied are titanium dioxide, alumina, zinc oxide, carbon black, and carbon nanotubes, and "nano-C60". Due to quantum size effects and large surface area to volume ratio, nanomaterials are highly reactive which potentially lead to toxicity as a result of their interactions with biological systems and the environment. Some nanoparticles seem to be able to translocate from their site of deposition to distant sites such as the blood and the brain. This has resulted in a sea-change in how particle toxicology is viewed- instead of being confined to the lungs; nanoparticle toxicologists study the brain, blood, liver, skin and gut. Nanotoxicology has revolutionised particle toxicology and rejuvenated it. The purpose of developing the concept of nanotoxicity is to acknowledge and evaluate the hazards and risks of Nanomaterials and measure safety. This discipline will be the important contribution for the development of a sustainable and safe Nanotechnology. Nanotoxicology is an interdisciplinary approach involving Toxicology, materials science, medicine, molecular biology etc.

The types of toxic exposure of nanoparticle are as follows:

(1) Consumer exposure—usage of nanoparticle-containing personal care products, cosmetics, sunscreen preparations;

(2) Occupational exposure—for workers in nanomaterial manufacturing and research;

(3) Environmental exposure—the increasing concentrations of nanomaterials in groundwater and soil may produce a significant environmental risk.

Moreover, the toxicity of nanomaterials is broadly classified into two areas:

- Biological toxicity
- Environmental toxicity.

In biological systems, the nanostructured nanoparticles enter the body via six principle routes, i.e., oral, intravenous, dermal, subcutaneous, inhalation, and intraperitoneal. The particles enter the systemic circulation and are distributed to the various organs of the body and may remain structurally the same or be modified or metabolized. In cellular systems, they can cause biological toxicity by DNA damage or ROS generation. This leads to tissue damage, inflammation, cytotoxicity, and organ failure from the deposition of nanoparticles. And it also causes fibrosis and allergies. In environmental systems, non-degradable nanoparticles are ready to deposit in the groundwater, leading to production of environmental pollutants. Further, they may also be harmful to plants and microbes. The removal of nanoparticles from the environment is another challenging task in the management of nanotoxicity.

References

- Agrawal, Anju; Gopal, Krishna (2013). Biomonitoring of Water and Waste Water. Springer, India. pp. 135–147. doi:10.1007/978-81-322-0864-8_13. ISBN 9788132208631

- Environmental-toxicology, science: britannica.com, Retrieved 11 March 2018

- Liu, Fu-Jun; Wang, Jia-Sheng; Theodorakis, Chris W. (May 2006). "Thyrotoxicity of Sodium Arsenate, Sodium Perchlorate, and Their Mixture in ZebrafishDanio rerio". Environmental Science & Technology. 40 (10): 3429–3436. doi:10.1021/es052538g. ISSN 0013-936X

- Occupational-toxicology: thebts.org, Retrieved 21 May 2018

- "Lucille Farrier Stickel: Research Pioneer". National Wildlife Refuge System. United States Fish and Wildlife Service. March 7, 2014. Retrieved August 24, 2015

- Forensic-toxicology, pharmacology-toxicology-and-pharmaceutical-science: sciencedirect.com, Retrieved 14 April 2018

- Gagliano-Candela, R. and Aventaggiato, L. "The detection of toxic substances in entomological specimens." International Journal of Legal Medicine 114 (2001): 197-203

- Forensic-toxicology: exploreforensics.co.uk, Retrieved 24 May 2018

- van der Merwe, Deon (2014). "Chapter 31: Freshwater Cyanotoxins". In Gupta, Ramesh C. Biomarkers in Toxicology. Elsevier. pp. 539–548. doi:10.1016/b978-0-12-404630-6.00031-2. ISBN 9780124046306

- Entomotoxicology-alternative-matrices-forensic-toxicology: forensicmag.com, Retrieved 29 April 2018

- Goff, M. Lee and Lord, Wayne D. "Entomotoxicology- A New Area for Forensic Investigation." The American Journal of Forensic Medicine and Pathology 15 (1994): 51-57

- Nanotoxicology, pharmacology-toxicology-and-pharmaceutical-science: sciencedirect.com, Retrieved 13 April 2018

- "Dioxins and PCBs report shows drop in dietary exposure over last decade | European Food Safety Authority". www.efsa.europa.eu. Retrieved 2016-02-04

Chapter 3

Toxin and its Types

A toxin is a poisonous substance that is produced within the cells of living organisms. It can be peptides or small molecules, and vary in toxicity from minor complications to death. There are different classifications of toxins, such as microbial toxins, neuro-toxins, plant and animal toxins, which have been discussed in comprehensive detail in this chapter.

A toxin is a chemical substance that is capable of causing injury, illness, or death to an organism (poison) and that is produced by living cells or another organism. The term sometimes is used in a broader sense to refer to any substance that is poisonous to an organism, but generally the usage is limited to poisons produced via some biological function in nature, such as the bacterial proteins that cause tetanus and botulism. While the term is especially applied to substances of bacterial origin, many diverse taxa produce toxins, including dinoflagellates, fungi, plants, and animals.

Toxins are nearly always proteins that are capable of causing harm on contact or absorption with body tissues by interacting with biological macromolecules such as enzymes or cellular receptors. Toxins vary greatly in their severity, ranging from usually minor and acute (as in a bee sting) to almost immediately deadly (as in botulinum toxin).

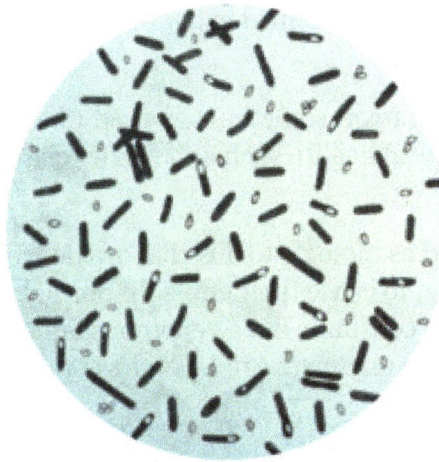

Clostridium botulinum, a bacteria that produces botulinum toxin, a neurotoxin protein that to humans is one of the most poisonous naturally-occurring substances in the world

Biotoxins vary greatly in purpose and mechanism, and they can be highly complex (the venom of the cone snail contains dozens of small proteins, each targeting a specific nerve channel or receptor), or a single, relatively small protein.

Human creativity has resulted in understanding about toxins and their mechanisms, and this knowledge has been employed in making effective insecticides, to improve the quality of human life, and in making vaccines and antidotes (such as antivenom to snake toxins). On the other hand, human creativity also has used this knowledge to create nerve agents designed for biological warfare and biological terrorism. For example, in 2001, powdered preparations of Bacillus anthracis spores were delivered to targets in the United States through the mail. Inhaling the weaponized spores can causes a form of quickly developing anthrax that is almost always fatal if not treated. Ricin, a toxin produced from the castor bean, has long been used as a weapon of terrorism, and is one for which there is no vaccine or antidote.

Terminology: Toxin, Poison, Venom

The term toxin comes from the Greek toxikon, meaning "(poison) for use on arrows." In the context of biology, poisons are substances that can cause damage, illness, or death to organisms, usually by chemical reaction or other activity on the molecular scale, when a sufficient quantity is absorbed by an organism.

Toxin is a subcategory of poison, referring to a substance produced by a living organism. However, when used non-technically, the term "toxin" is often applied to any poisonous substance. Many non-technical and lifestyle journalists also follow this usage to refer to toxic substances in general, though some specialist journalists maintain the distinction that toxins are only those produced by living organisms. In the context of alternative medicine the term toxin often is used nonspecifically as well to refer to any substance claimed to cause ill health, ranging anywhere from trace amounts of pesticides to common food items like refined sugar or additives like artificial sweeteners and MSG.

In pop psychology, the term toxin sometimes is used to describe things that have an adverse effect on psychological health, such as a "toxic relationship," "toxic work environment," or "toxic shame."

Venoms usually are defined as biologic toxins that are delivered subcutaneously, such as injected by a bite or sting, to cause their effect. In normal usage, a poisonous organism is one that is harmful to consume, but a venomous organism uses poison to defend itself while still alive. A single organism can be both venomous and poisonous.

The derivative forms "toxic" and "poisonous" are synonymous.

A weakened version of a toxin is called a toxoid. Toxids have been treated chemically or by heat to limit their toxicity while still allowing them to stimulate the formation of antibodies.

Functions of Toxins

Biotoxins in nature have two primary functions:

- Predation or invasion of a host (bacterium, spider, snake, scorpion, jellyfish, wasp).
- Defense (bee, poison dart frog, deadly nightshade, honeybee, wasp).

For example, a toxin may be used in assisting bacterial invasion of a host's cells or tissues or to combat the defense system of the host. A spider may use toxin to paralyze a larger prey, or a snake may use to subdue its prey. On the other hand, a honeybee sting, while of little benefit to the honeybee itself (which usually dies as a result of part of the abdomen ripping lose with the stinger), can help in discouraging predation on the bees or their hive products.

Sometimes, however, action of a toxin on an organism may not correlate to any direct benefit to the organism producing the toxin, but be accidental damage.

Types of Organisms Producing Toxins

Numerous types of organisms produce toxins. Some well-known examples are listed below.

Bacteria

The term toxin is used especially in terms of poisonous substances produced by bacteria. Examples include cholera toxin from Vibrio cholera, tetanus toxin from Clostridium tetani, botulism toxin from Clostridium botulinum, and anthrax toxin from Bacillus anthracis.

Bacterial toxins can damage the cell wall of the host (e.g., alpha toxin of Clostridium perfringens), stop the manufacture of protein in host cells or degrade the proteins (e.g., exotoxin A of Pseudomonas aeruginosa or the protein degrading toxins of Clostridium botulinum), or stimulate an immune response in the host that is so strong as to damage the host (e.g., three different toxins of Staphylococcus aureus resulting in toxic shock syndrome.

Bacterial toxins are classified as either exotoxins or endotoxins. An exotoxin is a soluble protein excreted by a microorganism, including bacteria, fungi, algae, and protozoa. An exotoxin can cause damage to the host by destroying cells or disrupting normal cellular metabolism. Endotoxins are potentially toxic natural compounds found inside pathogens such as bacteria. Classically, an endotoxin is a toxin that, unlike an exotoxin, is not secreted in soluble form, but is a structural component in bacteria that is released mainly when bacteria are lysed. Of course, exotoxins also may be released if the cell is lysed.

Both gram positive and gram negative bacteria produce exotoxins, while endotoxins mainly are produced by gram negative bacteria.

Types of Exotoxins

Exotoxins can be categorized by their mode of action on target cells.

- Type I toxins: Toxins that act on the cell surface. Type I toxins bind to a receptor on the cell surface and stimulate intracellular signaling pathways. For example, "superantigens" produced by the strains of Staphylococcus aureus and Streptococcus pyogenes cause toxic shock syndrome.

- Type II toxins: Membrane damaging toxins. These toxins are designed primarily to disrupt the cellular membrane. Many type II exotoxins have hemolysin activity, which causes red blood cells to lyse in vitro.

- Type III toxins: Intracellular toxins. Intracellular toxins must be able to gain access to the cytoplasm of the target cell to exert their effects. Some bacteria deliver toxins directly from their cytoplasm to the cytoplasm of the target cell through a needle-like structure. The effector proteins injected by the type III secretion apparatus of Yersinia into target cells are one example. Another well-known group of intracellular toxins is the AB toxins. The 'B'-subunit attaches to target regions on cell membranes, allowing the the 'A'-subunit to enter through the membrane and stimulate enzymatic actions that affects internal cellular bio-mechanisms. The structure of these toxins allows for the development of specific vaccines and treatments. For example, certain compounds can be attached to the B unit, which the body learns to recognize, and which elicits an immune response. This allows the body to detect the harmful toxin if it is encountered later, and to eliminate it before it can cause harm to the host. Toxins of this type include cholera toxin, pertussis toxin, Shiga toxin, and heat-labile enterotoxin from E. coli.

- Toxins that damage the extracellular matrix. These toxins allow the further spread of bacteria and consequently deeper tissue infections. Examples are hyaluronidase and collagenase.

Exotoxins are susceptible to antibodies produced by the immune system, but many exotoxins are so toxic that they may be fatal to the host before the immune system has a chance to mount defenses against it.

Endotoxin Examples

The prototypical examples of endotoxin are lipopolysaccharide (LPS) or lipo-oligo-saccharide (LOS) found in the outer membrane of various gram-negative bacteria. The term LPS is often used interchangeably with endotoxin, owing to its historical discovery. In the 1800s, it became understood that bacteria could secrete toxins into their environment, which became broadly known as "exotoxin." The term endotoxin came from the discovery that portions of gram-negative bacteria themselves can cause toxicity, hence the name endotoxin. Studies of endotoxin over the next 50 years revealed that the effects of "endotoxin" was in fact due to lipopolysaccharide.

LPS consist of a polysaccharide (sugar) chain and a lipid moiety, known as lipid A, which is responsible for the toxic effects. The polysaccharide chain is highly variable among different bacteria. Humans are able to produce antibodies to endotoxins after exposure but these are generally directed at the polysaccharide chain and do not protect against a wide variety of endotoxins.

There are, however, endotoxins other than LPS. For example, delta endotoxin of Bacillus thuringiensis makes crystal-like inclusion bodies next to the endospore inside the bacteria. It is toxic to larvae of insects feeding on plants, but is harmless to humans (as we do not possess the enzymes and receptors necessary for its processing followed by toxicity).

The only known gram positive bacteria that produces endotoxin is Listeria monocytogenes.

Dinoflagellates

Dinoflagellates can produce toxic substances of danger to humans. For instance, one should avoid consuming mussels along the west coast of the United States during the warmer months. This is because dinoflagellates create elevated levels of toxins in the water that do not harm the mussels, but if consumed by humans can bring on illness. Usually the United States government monitors the levels of toxins throughout the year at fishing sites.

Fungi

Two species of mold—Aspergillus flavus and Aspergillus parasiticus—produce aflatoxin, which can contaminate potatoes afflicted by the mold. This can lead to serious and even fatal illness.

Plants

Many plants produce toxins designed to protect against insects and other animal consumers, or fungi.

The roots of the tobacco plant produce a substance called nicotine, which is stored mainly in the leaves. Nicotine is a powerful poison and seems to benefit the plant by protecting it from insects, working by attacking the junctions between the insects' nerve cells. Tobacco leaves are sometimes soaked or boiled and the water sprayed on other plants as an organic insecticide. Nicotine is also a deadly poison to humans. Two to four drops (pure nicotine is an oily liquid) are a fatal dose for an adult. Smoking and chewing tobacco results in a much smaller dose; however, people have died as a result of mistaking wild tobacco for an edible herb and boiling and eating a large quantity.

Poison ivy, poison hemlock, and nightshade are other plants that produce toxins that work against humans.

Ricin is found in the castor bean plant, and is the third most deadly toxin to humans known, after the toxins produced by Clostridium botulinum and Clostridium tetani. There is no known vaccine or antidote, and if exposed symptoms can appear within hours (nausea, muscle spasms, severe lung damage, and convulsion) and death from pulmonary failure within three days.

Animals

Many animals use toxins for predation or defense. Well known examples include pit vipers, such as rattlesnakes, that possess hemotoxins that target and destroy red blood cells and are transmitted through the bloodstream; the brown recluse or "fiddle back" spider that uses necrotoxins that cause death in the cells they encounter and destroy all types of tissues; and the black widow spider, most scorpions, the box jellyfish, elapid snakes, and the cone snail that use neurotoxins that primarily affect the nervous system of animals.

The puffer fish produces the deadly toxin tetrodotoxin in its liver and ovaries; it blocks nerve conduction.

Neurotoxin

Neurotoxin is a substance that alters the structure or function of the nervous system.

More than 1,000 chemicals are known to have neurotoxic effects in animals. The substances include a wide range of natural and human-made chemical compounds, from snake venom and pesticides to ethyl alcohol, heroin, and cocaine.

Characteristics

Many neurotoxins are of external origin, entering the body from environmental sources. Others, however, are endogenous, being produced and existing within the body. Examples of endogenous neurotoxins include the neurotransmitters nitric oxide and glutamate. Although both are important to cell communication in the nervous system, they can become toxic to neurons in high concentrations.

The extent to which a neurotoxin affects nerve function depends on the toxicity of the substance and on the individual's age and health (particularly renal and hepatic health) at the time of exposure. It also depends on the level and frequency of contact with a chemical; the same substance can have both toxic and therapeutic effects at different concentration levels. For instance, vitamin A and vitamin B6 are vital to a healthy diet; however, they become neurotoxic in large doses. Some puffer fish (and certain other aquatic and terrestrial animals) carry tetrodotoxin, an extremely potent inhibitor of voltage-gated sodium channels on neurons; nonetheless, fugu, a dish prepared from puffer species, is a traditional culinary delicacy in Japan. When inhaled, less than one microgram of botulinum toxin, a protein produced by the bacterium Clostridium botulinum, is lethal to humans; the toxin, in the form of Botox, however, is also used for a variety of medicinal and cosmetic purposes. Likewise, life-saving chemotherapy treatment and antipsychotic drugs can also have neurotoxic effects, although the benefits often outweigh the risks.

Physiological Effects

Neurotoxins are absorbed through inhalation, ingestion, skin contact, or injection and can have immediate or long-lasting impacts by causing neurons to malfunction or by disrupting interneuron communication. Slurred words or poor coordination due to toxic effects on neurons from alcohol consumption, for example, are temporary, whereas cognitive damage caused by lead exposure is irreversible. Certain neurotoxins are highly potent and have been developed into chemical weapons. The nerve agent sarin, for example, is an organophosphorous compound that is classified as a weapon of mass destruction; sarin gas can kill a person within 10 minutes of exposure.

Young and elderly persons are particularly vulnerable to neurotoxic chemicals. In elderly individuals, a decline in neural function associated with aging can limit the ability to cope with the effects of neurotoxins, particularly for those with compromised liver metabolism or impaired renal function, which are the primary routes of toxin clearance from the body. Prenatal, postnatal, and early childhood exposure to certain chemicals can result in permanent damage to the developing brain, causing functional deficits that become apparent immediately or that emerge later in life. Studies have shown that the placenta cannot prevent many toxins from passing from mother to fetus; moreover, chemicals can be transferred through breast milk.

Certain chemicals found in the environment and in common household items have been linked to behavioral and cognitive problems in children. In the early 2000s, increased exposure to some of those chemicals during fetal and early childhood development—as result of overall increases in the use of chemicals in food production and in consumer products—was blamed for the sharp rise in disorders such as autism and attention-deficit/hyperactivity disorder (ADHD) in children. Although genetic factors play a role, neurotoxins such as arsenic, chlorpyrifos, DDT (dichlorodiphenyltrichloroethane), ethyl alcohol, fluoride, lead, polybrominated diphenyl ethers (PBDEs), polychlorinated biphenyls (PCBs), manganese, mercury, and toluene are major contributors to the prevalence of neurobehavioral disorders.

The impact of some neurotoxins, such as lead and ethyl alcohol, are well-documented. Fetal alcohol syndrome, characterized by brain damage and developmental delays in children, has been known since the 1970s to occur in children born to alcoholic mothers. Concern about lead contamination dates to ancient Rome, where the malleable metal was used to line aqueducts. Roman author and civil engineer Vitruvius noted that "in casting lead, the fumes from it destroy the vigor of the blood."

In the modern era, there is significant concern about the combined effects of moderate and even low-level exposure to multiple neurotoxins; more research is needed, however, to determine the physiological significance of such exposures. In addition, thousands of other chemicals are suspected of having neurotoxic effects, though many remain untested.

Microbial Toxin

Toxins that are produced by microorganisms such as bacteria and fungi are called microbial toxins. Such toxins damage host tissues and tamper with the natural immune system thereby inducing diseases and infections. Bacterial toxins are the most potent toxins known. An example of this is Botulinum toxin (BX) or Botox, which is a neurotoxic protein that causes botulism. It is the most lethal toxin known, but has wide applications in the medical and cosmetic industry. Similarly, there exist massive potential applications of microbial toxins in medicine. Such toxicological researches can have significant implications in furthering the development of novel drugs for fighting cancer, combating microbial virulence and as targeted tools in cellular biology and neurobiology.

Cyanotoxin

Cyanotoxins are produced by cyanobacteria, formerly known as blue-green algae. While critical to water and soil in the environment, cyanobacteria that grow rapidly and excessively in lakes, reservoirs and other surface waters can produce toxic algal "blooms." These algal blooms are most common during late summer and fall months and can appear as thick blue, green or brown mats on the water. Cyanobacteria can cause unpleasant tastes and odors in water even after it is cleaned at a treatment plant, and in some cases, cyanobacteria can produce potentially harmful cyanotoxins. In August of 2014, a toxin-producing algal bloom in Lake Erie prompted the water utility to issue a "do not drink" advisory to 500,000 Toledo residents.

People can be exposed to cyanotoxins in several ways:

- Ingesting fish or shellfish from waters containing cyanotoxins;
- Making skin contact with water containing cyanotoxins through showering or swimming or otherwise;
- Inhaling or ingesting toxins in the air when swimming or recreating in waters when cyanotoxins are present;
- Consuming drinking water containing cyanotoxins.

Cyanobacteria blooms generally result from excess nutrient runoff into the drinking water supply, although there are other factors such as warmer water temperature that increase their growth as well. Nutrients, especially nitrogen and phosphorus, often are washed into rivers or other drinking water sources from agricultural operations and even from urban fertilizers.

Mycotoxin

Mycotoxins are toxic compounds that are naturally produced by certain types of molds (fungi). Molds that can produce mycotoxins grow on numerous foodstuffs such as cereals, dried fruits, nuts and spices. Mold growth can occur either before harvest or after harvest, during storage, on/in the food itself often under warm, damp and humid conditions. Most mycotoxins are chemically stable and survive food processing.

Several hundred different mycotoxins have been identified, but the most commonly observed mycotoxins that present a concern to human health and livestock include aflatoxins, ochratoxin A patulin, fumonisins, zearalenone and nivalenol/deoxynivalenol. Mycotoxins appear in the food chain as a result of mold infection of crops both before and after harvest. Exposure to mycotoxins can happen either directly by eating infected food or indirectly from animals that are fed contaminated feed, in particular from milk.

Major Groups of Mycotoxins

Aflatoxins are amongst the most poisonous mycotoxins and are produced by certain molds (*Aspergillus flavus* and *Aspergillus parasiticus*) which grow in soil, decaying vegetation, hay, and grains. Crops that are frequently affected by *Aspergillus* spp. include cereals (corn, sorghum, wheat and rice), oilseeds (soybean, peanut, sunflower and cotton seeds), spices (chili peppers, black pepper, coriander, turmeric and ginger) and tree nuts (pistachio, almond, walnut, coconut and Brazil nut). The toxins can also be found in the milk of animals that are fed contaminated feed, in the form of aflatoxin M1. Large doses of aflatoxins can lead to acute poisoning (aflatoxicosis) and can be life threatening, usually through damage to the liver. Aflatoxins have also been shown to be genotoxic, meaning they can damage DNA and cause cancer in animal species. There is also evidence that they can cause liver cancer in humans.

Ochratoxin A is produced by several species of *Aspergillus* and *Penicillium* and is a common food-contaminating mycotoxin. Contamination of food commodities, such as cereals and cereal products, coffee beans, dry vine fruits, wine and grape juice, spices and liquorice, occurs worldwide. Ochratoxin A is formed during the storage of crops and is known to cause a number of toxic effects in animal species. The most sensitive and notable effect is kidney damage, but the toxin may also have effects on fetal development and on the immune system. Contrary to the clear evidence of kidney toxicity and kidney cancer due to ochratoxin A exposure in animals, this association in humans is unclear, however effects on kidney have been demonstrated.

Patulin is a mycotoxin produced by a variety of molds, particularly *Aspergillus*, *Penicillium* and *Byssochlamys*. Often found in rotting apples and apple products, patulin can also occur in various mouldy fruits, grains and other foods. Major human dietary sources of patulin are apples and apple juice made from affected fruit. The acute symptoms in animals include liver, spleen and kidney damage and toxicity to the immune system. For humans, nausea, gastrointestinal disturbances and vomiting have been reported. Patulin is considered to be genotoxic however a carcinogenic potential has not been demonstrated yet.

Fusarium fungi are common to the soil and produce a range of different toxins, including trichothecenes such as deoxynivalenol (DON), nivalenol (NIV) and T-2 and HT-2 toxins, as well as zearalenone (ZEN) and fumonisins. The formation of the moulds and toxins occur on a variety of different cereal crops. Different fusarium toxins are associated with certain types of cereal. For example, both DON and ZEN are often associated with wheat, T-2 and HT-2 toxins with oats, and fumonisins with maize (corn). Trichothecenes can be acutely toxic to humans, causing rapid irritation to the skin or intestinal mucosa and lead to diarrhoea. Reported chronic effects in animals include suppression of the immune system. ZEN has been shown to have hormonal, estrogenic effects and can cause infertility at high intake levels, particularly in pigs. Fumonisins have been related to oesophageal cancer in humans, and to liver and kidney toxicity in animals.

Citrinin is a toxin that was first isolated from *Penicillium citrinum*, but has been identified in over a dozen species of *Penicillium* and several species of *Aspergillus*. Some of these species are used to produce human foodstuffs such as cheese (*Penicillium camemberti*), sake, miso, and soy sauce (*Aspergillus oryzae*). Citrinin is associated with yellowed rice disease in Japan and acts as a nephrotoxin in all animal species tested. Although it is associated with many human foods (wheat, rice, corn, barley, oats, rye, and food colored with Monascus pigment) its full significance for human health is unknown. Citrinin can also act synergistically with Ochratoxin A to depress RNA synthesis in murine kidneys.

Ergot Alkaloids are compounds produced as a toxic mixture of alkaloids in the sclerotia of species of *Claviceps*, which are common pathogens of various grass species. The ingestion of ergot sclerotia from infected cereals, commonly in the form of bread produced from contaminated flour, cause ergotism, the human disease historically known as St. Anthony's Fire. There are two forms of ergotism: gangrenous, affecting blood supply to extremities, and convulsive, affecting the central nervous system. Modern methods of grain cleaning have significantly reduced ergotism as a human disease; however, it is still an important veterinary problem. Ergot alkaloids have been used pharmaceutically.

Occurrence

Although various wild mushrooms contain an assortment of poisons that are definitely fungal metabolites causing noteworthy health problems for humans, they are rather arbitrarily excluded from discussions of mycotoxicology. In such cases the distinction is based on the size of the producing fungus and human intention. Mycotoxin exposure is almost always accidental whereas with mushrooms improper identification and ingestion causing mushroom poisoning is commonly the case. Ingestion of misidentified mushrooms containing mycotoxins may result in hallucinations. The cyclopeptide-producing *Amanita phalloides* is well known for its toxic potential and is responsible for approximately 90% of all mushroom fatalities. The other primary mycotoxin groups found in mushrooms include: orellanine, monomethylhydrazine, disulfiram-like, hallucinogenic indoles, muscarinic, isoxazole, and gastrointestinal (GI)-specific irritants. The bulk of this article is about mycotoxins that are found in microfungi other than poisons from mushrooms or macroscopic fungi.

Indoor Environments

Buildings are another source of mycotoxins and people living or working in areas with mold increase their chances of adverse health effects. Molds growing in buildings can be divided into three groups — primary, secondary, and tertiary colonizers. Each group is categorized by the ability to grow at a certain water activity requirement. It has become difficult to identify mycotoxin production by indoor molds for many variables, such as:

1. They may be masked as derivatives

2. They are poorly documented

3. The fact that they are likely to produce different metabolites on building materials.

Some of the mycotoxins in the indoor environment are produced by *Alternaria*, *Aspergillus* (multiple forms), *Penicillium*, and *Stachybotrys*. *Stachybotrys chartarum* contains a higher number of mycotoxins than other molds grown in the indoor environment and has been associated with allergies and respiratory inflammation. The infestation of *S. chartarum* in buildings containing gypsum board, as well as on ceiling tiles, is very common and has recently become a more recognized problem. When gypsum board has been repeatedly introduced to moisture, *S. chartarum* grows readily on its cellulose face. This stresses the importance of moisture controls and ventilation within residential homes and other buildings. The negative health effects of mycotoxins are a function of the concentration, the duration of exposure and the subject's sensitivities. The concentrations experienced in a normal home, office or school are often too low to trigger a health response in occupants.

In the 1990s, public concern over mycotoxins increased following multimillion-dollar toxic mold settlements. The lawsuits took place after a study by the Center for Disease Control (CDC) in Cleveland, Ohio, reported an association between mycotoxins from *Stachybotrys* spores and pulmonary hemorrhage in infants. However, in 2000, based on internal and external reviews of their data, the CDC concluded that because of flaws in their methods, the association was not proven. *Stachybotrys* spores in animal studies have been shown to cause lung hemorrhaging, but only at very high concentrations.

One study by the Center of Integrative Toxicology at Michigan State University investigated the causes of Damp Building Related Illness (DBRI). They found that *Stachybotrys* is possibly an important contributing factor to DBRI. So far animal models indicate that airway exposure to *S. chartarum* can evoke allergic sensitization, inflammation, and cytotoxicity in the upper and lower respiratory tracts. Trichothecene toxicity appears to be an underlying cause of many of these adverse effects. Recent findings indicate that lower doses (studies usually involve high doses) can cause these symptoms.

Some toxicologists have used the Concentration of No Toxicological Concern (CoNTC) measure to represent the airborne concentration of mycotoxins that are expected to cause no hazard to humans (exposed continuously throughout a 70–yr lifetime). The resulting data of several studies have thus far demonstrated that common exposures to airborne mycotoxins in the built indoor environment are below the CoNTC, however agricultural environments have potential to produce levels greater than the CoNTC.

In Food

Mycotoxins can appear in the food chain as a result of fungal infection of crops, either by being eaten directly by humans or by being used as livestock feed.

In 2004 in Kenya, 125 people died and nearly 200 others were treated after eating aflatoxin-contaminated maize. The deaths were mainly associated with homegrown maize that had not been treated with fungicides or properly dried before storage. Due to food shortages at the time, farmers may have been harvesting maize earlier than normal to prevent thefts from their fields, so that the grain had not fully matured and was more susceptible to infection.

Spices are susceptible substrate for growth of mycotoxigenic fungi and mycotoxin production. Red chilli, black pepper, and dry ginger were found to be the most contaminated spices.

In Animal Food

There were outbreaks of dog food containing aflatoxin in North America in late 2005 and early 2006, and again in late 2011.

Mycotoxins in animal fodder, particularly silage, can decrease the performance of farm animals and potentially kill them. Several mycotoxins reduce milk yield when ingested by dairy cattle.

In Dietary Supplements

Contamination of medicinal plants with mycotoxins can contribute to adverse human health problems and therefore represents a special hazard. Numerous natural occurrences of mycotoxins in medicinal plants and herbal medicines have been reported from various countries including Spain, China, Germany, India, Turkey and from the Middle East. In a 2015 analysis of plant-based dietary supplements, the highest mycotoxin concentrations were found in milk thistle-based supplements, at up to 37 mg/kg.

Health Effects

Some of the health effects found in animals and humans include death, identifiable diseases or health problems, weakened immune systems without specificity to a toxin, and as allergens or irritants. Some mycotoxins are harmful to other micro-organisms such as other fungi or even bacteria; penicillin is one example. It has been suggested that mycotoxins in stored animal feed are the cause of rare phenotypical sex changes in hens that causes them to look and act male.

In Humans

Mycotoxicosis is the term used for poisoning associated with exposures to mycotoxins. Mycotoxins have the potential for both acute and chronic health effects via ingestion, skin contact, inhalation, and entering the blood stream and lymphatic system. They inhibit protein synthesis, damage macrophage systems, inhibit particle clearance of the lung, and increase sensitivity to bacterial endotoxin.

The symptoms of mycotoxicosis depend on the type of mycotoxin; the concentration and length of exposure; as well as age, health, and sex of the exposed individual. The synergistic effects associated with several other factors such as genetics, diet, and interactions with other toxins have been poorly studied. Therefore, it is possible that vitamin deficiency, caloric deprivation, alcohol abuse, and infectious disease status can all have compounded effects with mycotoxins.

Mitigation

Mycotoxins greatly resist decomposition or being broken down in digestion, so they remain in the food chain in meat and dairy products. Even temperature treatments, such as cooking and freezing, do not destroy some mycotoxins.

Removal

In the feed and food industry it has become common practice to add mycotoxin binding agents such as montmorillonite or bentonite clay in order to effectively adsorb the mycotoxins. To reverse the adverse effects of mycotoxins, the following criteria are used to evaluate the functionality of any binding additive:

- Efficacy of active component verified by scientific data

- A low effective inclusion rate

- Stability over a wide pH range

- High capacity to absorb high concentrations of mycotoxins

- High affinity to absorb low concentrations of mycotoxins

- Affirmation of chemical interaction between mycotoxin and adsorbent

- Proven *in vivo* data with all major mycotoxins

- Non-toxic, environmentally friendly component

Since not all mycotoxins can be bound to such agents, the latest approach to mycotoxin control is mycotoxin deactivation. By means of enzymes (esterase, de-epoxidase), yeast (*Trichosporon mycotoxinvorans*), or bacterial strains (Eubacterium BBSH 797 developed by Biomin), mycotoxins can be reduced during pre-harvesting contamination. Other removal methods include physical separation, washing, milling, nixtamalization, heat-treatment, radiation, extraction with solvents, and the use of chemical or biological agents. Irradiation methods have proven to be effective treatment against mold growth and toxin production.

Regulations

Many international agencies are trying to achieve universal standardization of regulatory limits for mycotoxins. Currently, over 100 countries have regulations regarding mycotoxins in the feed industry, in which 13 mycotoxins or groups of mycotoxins are of concern. The process of assessing a need for mycotoxin regulation includes a wide array of in-laboratory testing that includes extracting, clean-up and separation techniques. Most official regulations and control methods are based on high-performance liquid techniques (e.g., HPLC) through international bodies. It is implied that any regulations regarding these toxins will be in co-ordinance with any other countries with which a trade agreement exists. Many of the standards for the method performance analysis for mycotoxins is set by the European Committee for Standardization (CEN). However, one must take note that scientific risk assessment is commonly influenced by culture and politics, which, in turn, will affect trade regulations of mycotoxins.

Food-based mycotoxins were studied extensively worldwide throughout the 20th century. In Europe, statutory levels of a range of mycotoxins permitted in food and animal feed are set by a range of European directives and EC regulations. The U.S. Food and Drug Administration has regulated and enforced limits on concentrations of mycotoxins in foods and feed industries since 1985. It is through various compliance programs that the FDA monitors these industries to guarantee that mycotoxins are kept at a practical level. These compliance programs sample food products including peanuts and peanut products, tree nuts, corn and corn products, cottonseed, and milk. There is still a lack of sufficient surveillance data on some mycotoxins that occur in the U.S.

Use in Fiction

A fictional use of a mycotoxin occurs in William Gibson's seminal novel *Neuromancer*. A "Russian war-time mycotoxin" is administered to Case, the novel's protagonist.

Exotoxin

Exotoxins are usually secreted by bacteria and act at a site removed from bacterial growth. However, in some cases, exotoxins are only released by lysis of the bacterial cell. Exotoxins are usually proteins, minimally polypeptides, that act enzymatically or through direct action with host cells and stimulate a variety of host responses. Most exotoxins act at tissue sites remote from the original point of bacterial invasion or growth. However, some bacterial exotoxins act at the site of pathogen colonization and may play a role in invasion.

Bacterial Protein Toxins

Exotoxins are usually secreted by living bacteria during exponential growth. The production of the toxin is generally specific to a particular bacterial species that produces the disease associated with the toxin (e.g. only *Clostridium tetani*produces tetanus toxin; only *Corynebacterium diphtheriae* produces the diphtheria toxin). Usually, virulent strains of the bacterium produce the toxin while nonvirulent strains do not, and the toxin is the major determinant of virulence (e.g. tetanus and diphtheria). At one time, it was thought that exotoxin production was limited mainly to Gram-positive bacteria, but clearly both Gram-positive and Gram-negative bacteria produce soluble protein toxins.

Bacterial protein toxins are the most powerful human poisons known and retain high activity at very high dilutions. The lethality of the most potent bacterial exotoxins is compared to the lethality of strychnine, snake venom, and endotoxin in table below.

Table: Lethality of Bacterial Protein Toxins

Toxin	Toxic Dose (mg)	Host	Lethal toxicity	compared with:	
			Strychnine	Endotoxin (LPS)	Snake Venom
Botulinum toxin	0.8×10^{-8}	Mouse	3×10^6	3×10^7	3×10^5
Tetanus toxin	4×10^{-8}	Mouse	1×10^6	1×10^7	1×10^5
Shiga toxin	2.3×10^{-6}	Rabbit	1×10^6	1×10^7	1×10^5
Diphtheria toxin	6×10^{-5}	Guinea pig	2×10^3	2×10^4	2×10^2

Usually the site of damage caused by an exotoxin indicates the location for activity of that toxin. Terms such as enterotoxin, neurotoxin, leukocidin or hemolysinare descriptive terms that indicate the target site of some well-defined protein toxins. A few bacterial toxins that obviously bring about the death of an animal are known simply as lethal toxins, and even though the tissues affected and the target site or substrate may be known, the precise mechanism by which death occurs is not clear (e.g. anthrax LF).

Some bacterial toxins are utilized as invasins because they act locally to promote bacterial invasion. Examples are extracellular enzymes that degrade tissue matrices or fibrin, allowing the bacteria to spread. This includes collagenase, hyaluronidase and streptokinase. Other toxins, also considered invasins, degrade membrane components, such as phospholipases and lecithinases. The pore-forming toxins that insert a pore into eucaryotic membranes are considered as invasins, as well.

Some protein toxins have very specific cytotoxic activity (i.e., they attack specific types of cells). For example, tetanus and botulinum toxins attack only neurons. But some toxins (as produced by staphylococci, streptococci, clostridia, etc.) have fairly broad cytotoxic activity and cause nonspecific death of various types of cells or damage to tissues, eventually resulting in necrosis. Toxins that are phospholipases act in this way. This is also true of pore-forming hemolysins and leukocidins.

Bacterial protein toxins are strongly antigenic. *In vivo*, specific antibody neutralizes the toxicity of these bacterial exotoxins (antitoxin). However, *in vitro,*specific antitoxin may not fully inhibit their activity. This suggests that the antigenic determinant of the toxin may be distinct from the active portion of the protein molecule. The degree of neutralization of the active site may depend on the distance from the antigenic site on the molecule. However, since the toxin is fully neutralized *in vivo*, this suggests that other host factors must play a role in toxin neutralization in nature.

Protein exotoxins are inherently unstable. In time they lose their toxic properties but retain their antigenic ones. This was first discovered by Ehrlich who coined the term "toxoid" for this product. Toxoids are detoxified toxins which retain their antigenicity and their immunizing capacity. The formation of toxoids can be accelerated by treating toxins with a variety of reagents including formalin, iodine, pepsin, ascorbic acid, ketones, etc. The mixture is maintained at 37 degrees at pH range 6 to 9 for several weeks. The resulting toxoids can be used for artificial immunization against diseases caused by pathogens where the primary determinant of bacterial virulence is toxin production. Toxoids are effective immunizing agents against diphtheria and tetanus that are part of the DPT (DTP) vaccine.

Biological Effects of Some Bacterial Exotoxins with Enzymatic Activity

- Cholera toxin (A-5B)

    ADP ribosylates eucaryotic adenylate cyclase Gs regulatory protein

Activates adenylate cyclase; increased level of intracellular cAMP promote secretion of fluid and electrolytes in intestinal epithelium leading to diarrhea

- Diphtheria toxin (A/B)

ADP ribosylates elongation factor 2

Inhibits protein synthesis in animal cells resulting in death of the cells

- Pertussis toxin (A-5B)

ADP ribosylates adenylate cyclase Gi regulatory protein

Blocks inhibition of adenylate cyclase; increased levels of cAMP affect hormone activity and reduce phagocytic activity

- E. coli heat-labile toxin LT (A-5B)

ADP ribosylates adenylate cyclase Gs regulatory protein. Similar or identical to cholera toxin

- Shiga toxin (A/5B)

Glycosidase cleavage of ribosomal RNA (cleaves a single Adenine base from the 28S rRNA)

Inactivates the mammalian 60S ribosomal subunit and leads to inhibition of protein synthesis and death of the susceptible cells; pathology is diarrhea, hemorrhagic colitis (HC) and hemolytic uremic syndrome (HUS)

- Pseudomonas Exotoxin A (A/B)

ADP ribosylates elongation factor-2 analogous to diphtheria toxin

Inhibits protein synthesis in susceptible cells, resulting in death of the cells

- Botulinum toxin (A/B)

Zn++ dependent protease acts on synaptobrevin at motor neuron ganglioside

Inhibits presynaptic acetylycholine release from peripheral cholinergic neurons resulting in flaccid paralysis

- Tetanus toxin (A/B)

Zn++ dependent protease acts on synaptobrevin in central nervous system

Inhibits neurotransmitter release from inhibitory neurons in the CNS resulting in spastic paralysis

- Anthrax toxin LF (A2+B)

Metallo protease that cleaves MAPKK (mitogen-activated protein kinase kinase) enzymes

Combined with the B subunit (PA), LF induces cytokine release and death of target cells or experimental animals

- Bordetella pertussis AC toxin (A/B) and Bacillus anthracis EF (A1+B)

Calmodulin-regulated adenylate cyclases that catalyze the formation of cyclic AMP from ATP in susceptible cells, as well as the formation of ion-permeable pores in cell membranes. Increases cAMP in phagocytes leading to inhibition of phagocytosis by neutrophils and macrophages; also causes hemolysis and leukolysis

- Staphylococcus aureus Exfoliatin B

Cleaves desmoglein 1, a cadherin found in desmosomes in the epidermis

(also a superantigen)

Separation of the stratum granulosum of the epidermis, between the living layers and the superficial dead layers.

Several bacterial toxins can act directly on the T cells and antigen-presenting cells of the immune system. Impairment of the immunologic functions of these cells by toxin can lead to human disease. One large family of toxins in this category are the so-called pyrogenic exotoxins produced by staphylococci and streptococci, whose biological activities include potent stimulation of the immune system, pyrogenicity, and enhancement of endotoxin shock.

Pyrogenic exotoxins are secreted toxins of 22 kDa to 30 kDa, and include staphylococcal enterotoxins serotypes A-E, G, and H; group A streptococcal pyrogenic exotoxins A-C; staphylococcal exfoliatin toxin; and staphylococcal TSST-1.

Endotoxin

Endotoxins are part of the outer membrane of the cell wall of Gram-negative bacteria. Endotoxin is invariably associated with Gram-negative bacteria whether the organisms are pathogenic or not. Although the term "endotoxin" is occasionally used to refer to any cell-associated bacterial toxin, in bacteriology it is properly reserved to refer to the lipopolysaccharide complex associated with the outer membrane of Gram-negative pathogens such as *Escherichia coli, Salmonella, Shigella, Pseudomonas, Neisseria, Haemophilus influenzae, Bordetella pertussis* and *Vibrio cholerae*.

The relationship of endotoxin (lipopolysaccharide) to the bacterial cell surface is illustrated in figure below:

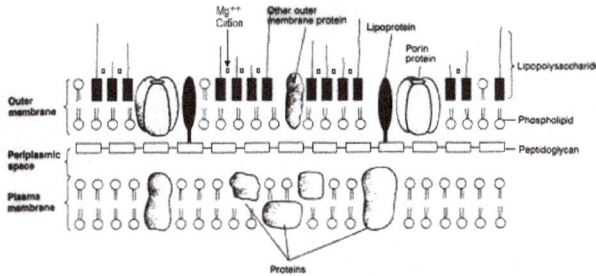

Figure: Structure of the cell envelope of a Gram-negative bacterium.

The biological activity of endotoxin is associated with the lipopolysaccharide (LPS). Toxicity is associated with the lipid component (Lipid A) and immunogenicity is associated with the polysaccharide components. The cell wall antigens (O antigens) of Gram-negative bacteria are components of LPS. LPS elicits a variety of inflammatory responses in an animal and it activates complement by the alternative (properdin) pathway, so it may be a part of the pathology of Gram-negative bacterial infections.

In vivo, Gram-negative bacteria probably release minute amounts of endotoxin while growing. This may be important in the stimulation of natural immunity. It is known that small amounts of endotoxin may be released in a soluble form by young cultures grown in the laboratory. But for the most part, endotoxins remain associated with the cell wall until disintegration of the organisms. *In vivo*, this results from autolysis, external lysis mediated by complement and lysozyme, and phagocytic digestion of bacterial cells.

Compared to the classic exotoxins of bacteria, endotoxins are less potent and less specific in their action, since they do not act enzymatically. Endotoxins are heat stable (boiling for 30 minutes does not destabilize endotoxin), but certain powerful oxidizing agents such as superoxide, peroxide and hypochlorite, have been reported to neutralize them. Endotoxins, although antigenic, cannot be converted to toxoids. A comparison of the properties of bacterial endotoxins and classic exotoxins is shown in table.

Table: Characteristics of bacterial endotoxins and classic exotoxin

PROPERTY	ENDOTOXIN	EXOTOXIN
Chemical Nature	Lipopolysaccharide (mw = 10kDa)	Protein (mw = 50-1000kDa)
Relationship To Cell	Part of outer membrane	Extracellular, diffusible
Denatured By Boiling	No	Usually
Antigenic	Yes	Yes
Form Toxoid	No	Yes
Potency	Relatively low (>100ug)	Relatively high (1 ug)
Specificity	Low degree	High degree
Enzymatic Activity	No	Often
Pyrogenicity	Yes	Occasionally

The Role of LPS in the Outer Membrane of Gram-negative Bacteria

The function of the outer membrane of Gram-negative bacteria is to act as a protective permeability barrier. The outer membrane is impermeable to large molecules and hydrophobic compounds from the environment. LPS is essential to the function of the outer membrane. First, it establishes a permeability barrier that is permeable only to low molecular weight, hydrophilic molecules. In the *E. coli* the ompF and ompC porins exclude passage of all hydrophobic molecules and any hydrophilic molecules greater than a molecular weight of about 700 daltons. This prevents penetration of the bacteria by bile salts and other toxic molecules from the GI tract. It also a barrier to lysozyme and many antimicrobial agents. Second, in an animal host, it may impede destruction of the bacterial cells by serum components and phagocytic cells. Third, LPS may play a role as an adhesin used in colonization of the host. Lastly, variations in LPS structure provide for the existence of different antigenic strains of a pathogen that may be able to bypass a previous immunological response to a related strain.

Chemical Nature of Endotoxin

Most of the work on the chemical structure of endotoxin has been done with species of *Salmonella* and *E. coli*. LPS can be extracted from whole cells by treatment with 45% phenol at 90°. Mild hydrolysis of LPS yields Lipid A plus polysaccharide.

Lipopolysaccharides are complex amphiphilic molecules with a mw of about 10kDa, that vary widely in chemical composition both between and among bacterial species The general architecture of LPS is shown in figure. The general structure of *Salmonella* LPS is shown in figure and the complete structure of *Salmonella*lipid A is illustrated in figure.

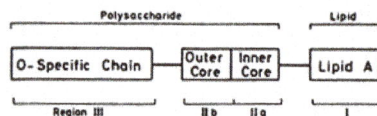

Figure: General architecture of Lipopolysaccharide

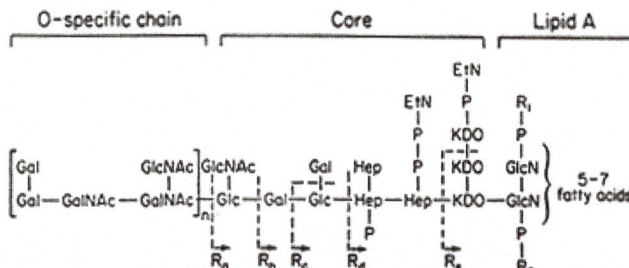

Figure: General Structure of *Salmonella* LPS.

Glc = glucose; GlcNac = N-acetyl- glucosamine; Gal = galactose; Hep = heptose; P = phosphate; Etn = ethanolamine; R1 and R2 = phoshoethanolamine or aminoarabinose. Ra to Re indicate incomplete forms of LPS. The Rd2 phenotype (not shown) would have only a single heptose unit. The Rc, Rd2, and Rd1 mutants lack the phosphate group attached to Hep.

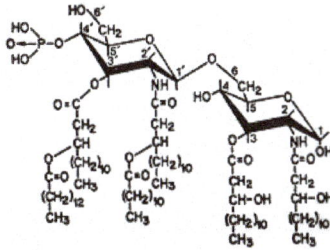

Figure: Complete structure of the Lipid A Moiety of LPS of *S. typhimurium*, *S. minnesota*, and *E. coli*

LPS consists of three components or regions: Lipid A, an R polysaccharide and an O polysaccharide.

Region I. Lipid A is the lipid component of LPS. It contains the hydrophobic, membrane-anchoring region of LPS. Lipid A consists of a phosphorylated N-acetylglucosamine (NAG) dimer with 6 or 7 fatty acids (FA) attached. Usually 6 FA are found. All FA in Lipid A are saturated. Some FA are attached directly to the NAG dimer and others are esterified to the 3-hydroxy fatty acids that are characteristically present. The structure of Lipid A is highly conserved among Gram-negative bacteria. Among *Enterobacteriaceae* Lipid A is virtually constant.

The primary structure of Lipid A has been elucidated and Lipid A has been chemically synthesized. Its biological activity appears to depend on a peculiar conformation that is determined by the glucosamine disaccharide, the PO_4 groups, the acyl chains, and also the KDO-containing inner core.

Region II. Core (R) antigen or R polysaccharide is attached to the 6 position of one NAG. The R antigen consists of a short chain of sugars. For example: KDO - Hep - Hep - Glu - Gal - Glu - GluNAc - Two unusual sugars, heptose and 2-keto-3-deoxyoctonoic acid (KDO), are usually present, in the core polysaccharide. KDO is unique and invariably present in LPS and so it has been used as an indicator in assays for LPS (endotoxin).

With minor variations, the core polysaccharide is common to all members of a bacterial genus (e.g. *Salmonella*), but it is structurally distinct in other genera of Gram-negative bacteria. *Salmonella*, *Shigella* and *Escherichia* have similar but not identical cores.

Region III. Somatic (O) antigen or O polysaccharide is attached to the core polysaccharide. It consists of repeating oligosaccharide subunits made up of 3 - 5 sugars. The individual chains vary in length ranging up to 40 repeat units. The O polysaccharide is much longer than the core polysaccharide, and it maintains the hydrophilic domain of the LPS molecule. A major antigenic determinant (antibody-combining site) of the Gram-negative cell wall resides in the O polysaccharide.

Great variation occurs in the composition of the sugars in the O side chain between species and even strains of Gram-negative bacteria. At least 20 different sugars are known to occur and many of these sugars are characteristically unique dideoxyhexoses, which occur in nature only in Gram-negative cell walls. Variations in sugar content of the O polysaccharide contribute to the wide variety of antigenic types of *Salmonella* and *E. coli* and presumably other strains of Gram-negative species. Particular sugars in the structure, especially the terminal ones, confer immunological specificity of the O antigen, in addition to "smoothness" (colony morphology) of the strain. Loss of the O specific region by mutation results in the strain becoming a "rough" (colony morphology) or R strain.

The elucidation of the structure of LPS relied heavily on the availability of mutants each blocked at a particular step in LPS synthesis. The biosynthesis of LPS is strictly sequential. The core sugars are added sequentially to Lipid A by successive additions, and the O side chain is added last, one preassembled subunit at a time. The properties of mutants producing incomplete LPS molecules suggests the nature and biological functions performed by various parts of the LPS molecule.

In *E. coli* and *Salmonella*, loss of the O antigen results in partial loss of virulence, suggesting that this portion of LPS is important during a host-parasite interaction. It is known that such "rough" mutants are more susceptible to phagocytosis and serum bactericidal reactions.

Loss of the more proximal parts of the core, as in "deep rough" mutants makes the strains sensitive to a range of hydrophobic compounds, including antibiotics, detergents, bile salts and mutagens. This area contains a large number of charged groups and is thought to be important in maintaining the permeability properties of the outer membrane.

Mutants in the assembly of Lipid A cannot be isolated except as conditional lethal mutants and this region must therefore be essential for cell viability. The innermost region of LPS, consisting of Lipid A and three residues of KDO, appears to be essential for viability, presumably for assembling the outer membrane.

LPS and Virulence of Gram-negative Bacteria

Both Lipid A (the toxic component of LPS) and the polysaccharide side chains (the non-toxic but immunogenic portion of LPS) act as determinants of virulence in Gram-negative bacteria.

The O-polysaccharide and Virulence

Virulence, and the property of "smoothness", is associated with an intact O-polysaccharide, The involvement of the polysaccharide chain in virulence is shown by the fact that small changes in the sugar sequences in the side chains of LPS result in major changes in virulence. How is the polysaccharide side chains involved in the expression of virulence? There are a number of possibilities:

1. O-specific antigens could allow organisms to adhere specifically to certain tissues, especially epithelial tissues.

2. Smooth antigens probably allow resistance to phagocytes, since rough mutants are more readily engulfed and destroyed by phagocytes.

3. The hydrophilic O-polysaccharides could act as water-solubilizing carriers for toxic Lipid A. It is known that the exact structure of the polysaccharide can greatly influence water binding capacity at the cell surface.

4. The O-antigens could provide protection from damaging reactions with antibody and complement. Rough strains of Gram-negative bacteria derived from virulent strains are generally non virulent. Smooth strains have polysaccharide "whiskers" which bear O-antigens projecting from the cell surface. The O-antigens are the key targets for the action of host antibody and complement, but when the reaction takes place at the tips of the polysaccharide chains, a significant distance external to the general bacterial cell surface, complement fails to have its normal lytic effect. Such bacteria are virulent because of this resistance to immune forces of the host. If the projecting polysaccharide chains are shortened or removed, antibody reacts with antigens on the general bacterial surface, or very close to it, and complement can lyse the bacteria. This contributes to the loss of virulence in "rough" colonial strains.

5. The O-polysaccharide or O-antigen is the basis of antigenic variation among many important Gram-negative pathogens including *E. coli*, *Salmonella* and *Vibrio cholerae*. Antigenic variation guarantees the existence of multiple serotypes of the bacterium, so that it is afforded multiple opportunities to infect its host if it can bypass the immune response against a different serotype. Furthermore, even though the O-polysaccharides are strong antigens, they seldom elicit immune responses which give full protection to the host against secondary challenge with specific endotoxin.

Lipid A and Virulence

The physiological activities of LPS are mediated mainly by the Lipid A component of LPS. Lipid A is a powerful biological response modifier that can stimulate the mammalian immune system. During infectious disease caused by Gram-negative bacteria, endotoxins released from, or part of, multiplying cells have similar effects on animals and significantly contribute to the symptoms and pathology of the disease encountered.

Since Lipid A is embedded in the outer membrane of bacterial cells, it probably only exerts its toxic effects when released from multiplying cells in a soluble form, or when the bacteria are lysed as a result of autolysis, complement and the membrane attack complex (MAC), ingestion and killing by phagocytes, or killing with certain types of antibiotics.

The injection of living or killed Gram-negative cells or purified LPS into experimental animals causes a wide spectrum of nonspecific pathophysiological reactions, such as fever, changes in white blood cell counts, disseminated intravascular coagulation, hypotension, shock and death. Injection of fairly small doses of endotoxin results in death in most mammals. The sequence of events follows a regular pattern:

(1) Latent period;

(2) Physiological distress (diarrhea, prostration, shock);

(3) Death how soon death occurs varies on the dose of the endotoxin, route of administration, and species of animal. Animals vary in their susceptibility to endotoxin.

The mechanism is complex. In humans, LPS binds to a lipid binding protein (LBP) in the serum which transfers it to CD14 on the cell membrane, which in turn transfers it to another non-anchored protein, MD2, which associates with Toll-like receptor-4 (TLR4). This triggers the signaling cascade for macrophage/endothelial cells to secrete pro-inflammatory cytokines and nitric oxide that lead to characteristic "endotoxic shock". CD14 and TLR4 are present on several cells of the immunological system cells, including macrophages and dendritic cells. In monocytes and macrophages, three types of events are triggered during their interaction with LPS:

1. Production of cytokines, including IL-1, IL-6, IL-8, tumor necrosis factor (TNF) and platelet-activating factor. These, in turn, stimulate production of prostaglandins and leukotrienes. These are powerful mediators of inflammation and septic shock that accompanies endotoxin toxemia. LPS activates macrophages to enhanced phagocytosis and cytotoxicity. Macrophages are stimulated to produce and release lysosomal enzymes, IL-1 ("endogenous pyrogen"), and tumor necrosis factor (TNFalpha), as well as other cytokines and mediators.

2. Activation of the complement cascade: C3a and C5a cause histamine release (leading to vasodilation) and affect neutrophil chemotaxis and accumulation. The result is inflammation.

3. Activation of the coagulation cascade: Initial activation of Hageman factor (blood-clotting Factor XII) can activate several humoral systems resulting in:

   a. Coagulation: a blood clotting cascade that leads to coagulation, thrombosis, acute disseminated intravascular coagulation, which depletes platelets and various clotting factors resulting in internal bleeding.

   b. Activation of the complement alternative pathway (as above, which leads to inflammation).

    c.   plasmin activation which leads to fibrinolysis and hemorrhaging.

    d.   kinin activation releases bradykinins and other vasoactive peptides which causes hypotension.

The net effect is to induce inflammation, intravascular coagulation, hemorrhage and shock.

LPS also acts as a B cell mitogen, stimulating the polyclonal differentiation and multiplication of B-cells and the secretion of immunoglobulins, especially IgG and IgM.

Plant Toxins

Plant toxins are toxic secondary plant metabolites which naturally occur in food, feed, weeds and ornamental plants. The chemical diversity is tremendous.

Aromatic plants which are used as an ingredient in food (herbs, spices), scents and flavours (essential oils) or traditional herbal remedies are examples of products in which plant toxins abundantly occur. In herbal remedies a plant toxin can be the same substance as the ingredient to which the pharmaceutical effect is attributed. In this case the difference between the toxic and pharmaceutical effect is obviously the dose.

Some poisonings by toxic plants have highly visible consequences, while others remain silent for weeks or months. Where, in the absence of a drought or seasonal feed shortage, there has been a dramatic change in the condition of animals, producers should suspect that disease, including poisoning by toxic plants, may be present. If this is the case a veterinary investigation should be carried out.

With all diseases, nutritional deficiencies and poisonings by toxic plants assess the risk based on previous local district history.

Managing Poisoning Risk from Toxic Plants

- Toxic plants may include pastures species at certain growth stages, native species and garden plants.

- The relative toxicity of plants may vary according to season and the stage of plant growth:

    o   Wilting in dry conditions and rapid growth after rain can increase the toxicity of some plants.

    o   Applying fertilizer to promote lush growth may increase toxicity.

    o   Some plants may only be toxic when growing in particular soil types.

    o   Stressful growth conditions, such as drought or insect attack, may cause toxins to concentrate in a plant.

  o   Plant parts can vary in their relative toxicity.

- Herbicide treatments can increase the palatability of plants.

- When livestock are hungry they may gorge themselves on things that they would
  not normally eat. Do not introduce hungry livestock to areas when toxic plants
  are known to be growing.

- Livestock grazing in a particular area for extended periods may become accus-
  tomed to eating small amounts of toxic plant material. New mobs introduced to
  the same area will not have the same tolerance.

Phytotoxin

Phytotoxin is a poison in a plant. Abrin and ricin are examples of phytotoxins. Abrin
comes from the seeds of a plant called the rosary pea or jequirity pea and ricin from the
seeds of the castor bean plant.

Phycotoxin

Phycotoxins are natural metabolites produced by micro-algae. Through accumulation
in the food chain, these toxins may concentrate in different marine organisms, including filter-feeding bivalves, burrowing and grazing organisms, herbivorous and predatory fish. Human poisoning due to ingestion of seafood contaminated by phycotoxins has
occurred in the past, and harmful algal blooms (HABs) are naturally occurring events.

Phycotoxins are small to medium-sized natural products and belong to many different groups of chemical compounds. The molecular mass ranges from ≈300 to over
3000 Da, and the compound classes represented include amino acids, alkaloids and
polyketides. Each compound group typically has several main compounds based on the
same or similar structure. However, most groups also have several analogues, which are
either produced by the algae or through metabolism in fish or shellfish or other marine
organisms. The different phycotoxins have distinct molecular mechanisms of action.
Saxitoxins, ciguatoxins, brevetoxins, gambierol, palytoxins, domoic acid, and, perhaps,
cyclic imines, alter different ion channels and pumps at the level of the cell membrane.
The normal functioning of neuronal and other excitable tissues is primarily perturbed
by these mechanisms, leading to adverse effects in humans. Okadaic acid and related
compounds inhibit serine/threonine phosphoprotein phosphatases, and disrupt major
mechanisms controlling cellular functions. Pectenotoxins bind to actin filaments, and
alter cellular cytoskeleton. The precise mechanisms of action of yessotoxins and azaspi-
racids, in turn, are still undetermined. The route of human exposure to phycotoxins
is usually oral, although living systems may become exposed to phycotoxins through
other routes. Based on recorded symptoms, the major poisonings recognized so far
include paralytic, neurotoxic, amnesic, diarrheic shellfish poisonings, ciguatera, as well
as palytoxin and azaspiracid poisonings.

Potential Ecological Effects

Anti-grazing Effects

Phycotoxins may prevent grazing by several mechanisms: grazer death, infertility, or deterrence. Some evidence of anti-grazing effects:

1. Teegarden found that three different species of copepods were able to distinguish between a saxitoxin-producing Alexandrium sp. and morphologically similar non-toxigenic Alexandrium sp. by chemosensory means. These three different copepod species grazed predominantly on the non-toxigenic Alexandrium spp. and avoided the saxitoxin-producer. However, the effect of saxitoxin deterrence varied per copepod species. This implies that saxitoxin producing Alexandrium sp. have an advantage over non-toxigenic dinoflagellates.

2. Miralto reported a low hatching success of eggs laid by copepods that fed on diatoms containing polyunsaturated aldehydes. When ingested by copepods, these aldehydes appear to arrest embryonic development. This has the potential to decrease the future population of copepods and promote the survival of copepods which do not eat as many diatoms.

Anti-microbial Effects

Phycotoxins production may be useful to ward off parasitic or algicidal heterotrophic bacteria. Some evidence of anti-microbial effects:

1. Bates was able to enhance domoic acid production in Pseudo-nitzschia multseries with the re-introduction of bacteria. Additionally, P. multiseries cultures which were completely axenic (bacteria-free), produce less domoic acid than P. multiseries cultures which have contained bacteria for several generations.

2. Sieburth found acrylic acid inhibited gut microflora in penguins. High concentrations of acrylic acid were ingested by penguins via their euphasid diet, which had been feeding on Phaeocystis. The antimicrobial effect of acrylic acid was verified by Slezak who concluded that acrylic acid will inhibit bacterial production in situations where phytoplankton form aggregates (i.e. marine snow or Phaeocystis blooms). However, acrylic acid production may also serve to keep bacteria away from the phytoplankton in more dilute concentrations.

Competitive Effects

Since many different species of phytoplankton compete for a limited number of nutrients (see Paradox of the Plankton), it's possible that phycotoxin production is used as a method to either kill competitors or to keep other phytoplankton out of the producer's nutrients space. Some evidence of competitive effects:

1. Graneli showed that Prymnesium spp. will produce phycotoxins which kill competitors under nitrogen or phosphorus limitation.

2. Fistarol found that Alexandrium spp. produce toxins which decrease the growth rate of other phytoplankton and change community composition.

3. Prince showed that chemical exudates from the dinoflagellate Karenia brevis decreased the growth rate and sometimes killed competitor species by decreasing their photosynthetic efficiency and increasing membrane permeability.

Types of Toxins

Excreted Toxins

Excreted toxins may help deter predators and bacteria which are drawn in by phytoplankton waste products. Phytoplankton are known to excrete waste metabolites into the surrounding environment. This is a potential source of reduced nutrients and carbon for bacteria and may act as a signal for predators which can detect and follow kairomone gradients in their environment. Excreted toxins would seem most advantageous to the individual cell in their ability to keep predators and/or parasitic and algicidal bacteria at a distance. However, continuous toxin production and excretion carries a metabolic cost. For excreted toxins to be effective, they must have a low molecular weight to rapidly diffuse in the marine environment and to be energetically cheap to produce. However, excreted toxins may not actually repel larger motile predators because molecular diffusivity is slow and turbulence at the millimeter scale is large in water. Excreted phycotoxins may act as repellents if their signal registers at the same speed as other signals that potential grazers can detect (kairomones), assuming both are encountered by a predator at the same time. Additionally, excreted toxins may be effective method of keeping harmful bacteria and other phytoplankton competitors outside of the phycotoxin producer's microzone of nutrients.

Contact Toxins

Contact toxins are effective if they impact the grazer or harmful bacterium immediately after contact with the phytoplankton producer. These toxins are located at the cell surface and are typically classified as glycoproteins, glycolipids, or polypeptides. These toxins would have to be highly specific to their target receptors to be effective.

Post-ingestion Toxins

In order for these types of toxins to take effect, post-ingestion toxin producers have to be consumed by a grazer. Post-ingestion toxins, also known as suicide toxins, are not beneficial to individual cells because unlike terrestrial plants, phytoplankton do not have sacrificial tissue. However, if internal toxins do result in the death, decrease growth rate, infertility, or deterrence of a predator the remaining representatives of the plankton community may benefit. Community defense is most beneficial in a clonal population where toxigenic species are abundant, for example during a monospecific phytoplankton bloom.

Chemical Defense Signal Mechanisms

Table modified from Wolfe:

	Excreted toxins	Contact toxins	Post-ingestion toxins
Molecules	Small molecules but varied structures; organic and amino acids, sugars, short-chain lipids and derivatives	Glycoproteins, glycolipids, polypeptides	Varied: toxicants or toxins
Toxin properties	Aqueous solubility, diffusivity, lability, toxicity	Specificity, toxicity	Toxicity or concentration
Toxin location	Aqueous environment	Cell surface	Cell interior
Effect or mode of action	Negative kinesis/taxis: repellent	Release following capture: deterrent	Subsequent inhibition of feeding or toxicity reduced digestibility or growth efficiency
Benefit level	Individual or population, including competitors	Individual	Genetically similar population

Detection Methods

It is technically difficult to identify and characterize a metabolite that is produced in low concentrations and is secreted into a fluid that contains a diversity of other metabolites. Allelopathy is very difficult to observe in the field (with the exception of harmful algal blooms) because phycotoxin production may be induced by a variety of environmental factors and may create a cascade of biotic and physical events, which are difficult to separate from direct allelopathic effects of one species on another. There are six points (similar in logic to Koch's postulates) that must be established to rigorously prove that one species is chemically inhibiting another in an ecological system.

1. A pattern of inhibition of one species by another must be shown.

2. The putative aggressor species must produce a toxin.

3. There must be a mode of toxin release from the species into the environment.

4. There must be a mode of toxin transport and accumulation in the environment.

5. The afflicted species must have some means of toxin uptake.

6. The observed pattern of inhibition cannot be explained solely by physical factors or other biotic factors, especially competition and herbivory:

- Concentrations which impact the target species must be environmentally realistic given rates of transport and diffusion in the aquatic environment.

Few (if any) studies on phytoplankton toxins have attempted to rigorously meet all of these criteria. All methods of detecting phycotoxins involve extraction of the candidate toxin from a phytoplankton culture; therefore, it's important to determine whether the toxin is secreted into the media or stored in the phytoplankton cell. It's also important to know whether the target organism must be present to induce toxin synthesis.

Most commonly, the presence of a phycotoxin is verified by bioassay-guided fractionation. The sample must be fractionated, or separated from the other metabolites and chemicals in the media using chromatography. These different fractions may be then tested on the target species to determine which sample causes the expected allelopathic symptom(s). This approach is useful for rapidly isolating an allelochemical whose structure is not known. However, bioassays have the potential to generate false positives. This may occur if the bioassay is not controlled properly. For example, in a mixed batch culture the target species may die or have reduced growth rates due to competition for nutrients, dissolved inorganic carbon, or pH levels which are too low for the target species.

Developments in genomics, transcriptomics, proteomics, and metabolomics are now yielding large volumes of biochemical data. "Metabolic profiling" allows for comparison between biologically active and inactive samples and identification of compounds present at low concentrations using mass spectrometry. These samples may then be compared by principal component analysis. Characterization of the compounds present in the active sample (but not in the inactive sample) may then be identified and characterized using standard methods in mass spectroscopy. Isotope labeling may also be used to identify the pathways used in phycotoxin biosynthesis.

Animal Toxins

Envenoming by animal toxins has fascinated humans. Rarely has a medical phenomenon had so much religious association, symbolism, anecdotal communication, and provoked so much violent professional disagreement. Animal toxins have made a significant contribution to enhancing knowledge in human physiology and pharmacology. Information on the nature and mechanism of action of these toxins has enabled a more scientific approach to the treatment of their intoxications. Early and specific therapy is frequently required after envenoming and often includes life support and maintenance of vital functions by mechanical ventilation, i.v. fluid and drug therapy.

Animal toxins produce a wide range of physiological and pharmacological disturbances. Disorders of function at the neuromuscular junction are of particular interest and most intoxications require close monitoring and some form of intensive care. The role of toxins in the advancement of knowledge of human function is undeniable and further studies may prove invaluable in developing new drugs and techniques.

Venom

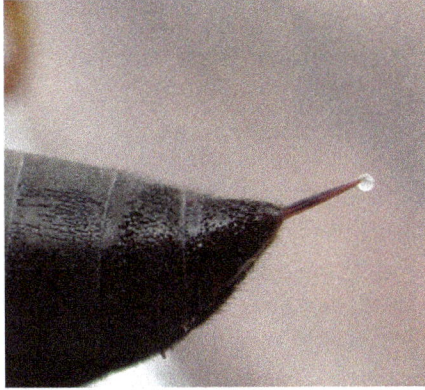

Wasp sting, with droplet of venom.

Venom is any of a variety of toxins produced by certain animals (such as snakes, stingrays, spiders, and scorpions) for mechanical delivery (injection) into another organism, usually by a bite, sting, or spine, causing injury, illness, or death in the victim. Venom typically is used for the purpose of defense and predation, although some venoms may provide an additional digestive function.

Venom is differentiated from the more generic term, poison, in that venom is necessarily produced by living cells and involves some delivery system or apparatus for introduction into the tissues and blood stream of another organism to cause its effect. Poison, on the other hand, is not necessarily produced by living cells and it may be absorbed by the body, such as through the skin or digestive system or even inhaled, rather than introduced by mechanical means by another organism. And poison tends to be widely distributed throughout the tissues of an animal, while venoms lack free distribution throughout the body and are produced, stored, and delivered by a very specific set of organs. Examples of venom delivery systems include injection through the hollow, hypodermic-like fangs of a rattlesnake or cobra, the barbed stinger of a honeybee, the thin, hollow spines in fins of lionfish, or the triggering of nematocysts by box jellyfish. The spitting cobras can shoot venom from the mouth and a hit to the eyes of the victim can cause temporary shock and blindness.

Among animals widely known to use venom are snakes (such as elapids and vipers), lizards (such as the Gila monster), spiders, centipedes, scorpions, stinging insects (such as bees and wasps), fish (stingrays, scorpionfish, along with many others), jellyfish, mollusks (such as cone snails), and even some mammals (such as the male platypus or slow loris).

A large number of venoms can disrupt the human nervous, cardiovascular, and muscular systems. However, despite the dangers that many venomous animals pose, venom also offers potential benefits. Research into snake venom has produced a number of potential stroke and cancer treatment drugs, and the venom of gila monsters offers promise for diabetes treatment. Among the attractions of venoms are their often staggering potency compared to plant compounds used for medicinal purposes and their often high specificity of action.

Venoms usually are defined as toxins secreted by certain animals that utilize an apparatus or delivery system to inject them into another organism, such as delivered subcutaneously by a bite, sting, spine, or other means.

While venom is occasionally, but rarely, used as a synonym of the more generic poison, in general there is a distinction made between "venomous organism" and "poisonous organism." Venomous, as stated above, refers to animals that deliver (often, inject) venom into their prey when hunting or as a defense mechanism. Poisonous, on the other hand, describes plants or animals that are harmful when consumed or touched. A poison also tends to be distributed over a large part of the body of the organism producing it, while venom is typically produced, stored, and delivered in organs specialized for this purpose and not distributed freely in other tissues. Poisonous animals lack localized or specific apparatuses for production, storage, or delivery of poisons, with the entire body, or a large part of it, poisonous.

Plants can be poisonous, not venomous. Animals can be venomous, poisonous, or both venomous and poisonous. The slow loris, a primate, has poison-secreting patches on the inside of its elbows, which it is believed to smear on its young to prevent them from being eaten. However, it will also lick these patches, giving it a venomous bite. On the other hand, the hooded pitohui, a type of bird, is not venomous, but is poisonous, secreting a neurotoxin onto its skin and feathers. Most snakes are venomous, but poisonous snakes are also known to exist. Rhabdophis tigrinus, a colubrid snake common in Japan, sequesters toxins from the toads it eats then secretes them from nuchal glands to ward off predators.

The distinction between poison and venom can be seen in the fact that snake venoms are generally not dangerous when ingested, as long as there are no lacerations inside the mouth or digestive tract; thus, they are not called poisons.

Examples of Venomous Animals

Venom is found among both vertebrates and invertebrates.

Among vertebrates, perhaps the most widely known venomous animals are the snakes, some species of which inject venom into their prey through hollow fangs. Over 2,000 species are known to be venomous. This number has recently increased greatly from a few hundred as research has revealed venom toxins in what previously were thought to be non-venomous snakes, and although these snakes have small quantities of venom and lack fangs, the venom still can be delivered via their sharp teeth. Although venom is found in several families of snakes, the two most well-known families of venomous snakes are Elapidae (including the cobras, mambas, and sea snakes) and Viperidae (vipers, such as rattlesnakes and puff adders).

The gila monster and bearded lizard are two lizards that have venom and venom delivery system, although similar venom toxins now have been found more widely in lizards.

Some believe venomous fish outnumber all other venomous vertebrates combined. Venom can also be found in some cartilaginous fish (stingrays, sharks, and chimaeras) and in teleost fish, including monognathus eels, catfishes, stonefishes, scorpionfishes, lionfishes, gurnard perches, rabbitfishes, surgeonfishes, scats, stargazers, weevers, carangids, saber-toothed blenny, and toadfish.

Although uncommon in mammals, two animals known to be venomous are the duck-billed platpus, which has a spur on the heel behind each leg, and the slow loris (Nycticebus), species of primates that can have a venomous bite from licking toxins produced from glands on the insides of the elbows. Some solenodons and shrews also are venomous.

Among invertebrates, animals widely known to use venom are spiders and centipedes, which inject venom through fangs; and scorpions and stinging insects, which inject venom with a sting (which, in insects such as bees and wasps, is a modified egg-laying device—the ovipositor). Many caterpillars have defensive venom glands associated with specialized bristles on the body, known as urticating hairs, and can be lethal to humans (for example, that of the Lonomia moth). The stinging hairs or spines of caterpillars of some moths may be hollow and connected to poison glands, with the venom flowing on contact. Various ants and true bugs produce venom as well. Other invertebrates that produce venom include jellyfish and cone snails. The box jellyfish is widely considered the most venomous creature in the world. Nematocysts, a venomous secretory product, are the cnidarians' main form of offense or defense. They function by a chemical or physical trigger that causes the specialized cell to eject a barbed and poisoned hook that can stick into, ensnare, or entangle prey or predators, killing or at least paralyzing its victim.

Snake Venom

Snake venoms are complex mixtures of proteins and are stored in glands at the back of the head. Typically, these glands open through ducts into grooved or hollow teeth in the upper jaw. These proteins can potentially be a mix of neurotoxins (which attack the nervous system), hemotoxins (which attack the circulatory system), cytotoxins, bungarotoxins, and many other toxins that affect the body in different ways. Almost all snake venom contains hyaluronidase, an enzyme that ensures rapid diffusion of the venom.

Venomous snakes that use hemotoxins usually have the fangs that secrete the venom in the front of their mouths, making it easier for them to inject the venom into their victims. Some snakes that use neurotoxins, such as the mangrove snake, have their fangs located in the back of their mouths, with the fangs curled backwards. This makes it both difficult for the snake to use its venom and for scientists to milk them. Vipers (Viperidae) have a single pair of long, hollow, venom-injecting fangs that can be folded back against the top of the mouth, tip inward, when the mouth is closed. Elapid snakes, however, such as cobras and kraits, are proteroglyphous, possessing hollow fangs that cannot be folded toward the roof of their mouths and cannot "stab" like a viper; they must actually bite the victim.

Snake bites cause a variety of symptoms including pain, swelling, tissue damage, low blood pressure, convulsions, and hemorrhaging (varying by species of snake).

Doctors treat victims of a venomous bite with antivenin, which can be created by dosing an animal such as a sheep, horse, goat, or rabbit with a small amount of the targeted venom. The immune system of the subject animal responds to the dose, producing antibodies to the venom's active molecule; the antibodies can then be harvested from the animal's blood and applied to treat envenomation in others. This treatment can be used effectively only a limited number of times for a given person, however, as that person will ultimately develop antibodies to neutralize the foreign animal antibodies injected into him (anti-antibody antibodies). Even if that person does not suffer a serious allergic reaction to the antivenom, his own immune system can destroy the antivenin before the antivenin can destroy the venom. Though most people never require even one treatment of antivenin in their lifetime, let alone several, people who work with snakes or other venomous animals may. Fortunately, these people may develop antibodies of their own against the venom of whatever animals they handle, and thereby are protected without the assistance of exogenous antibodies.

Spider Toxin

Spider toxins are a family of proteins produced by spiders which function as neurotoxins. The mechanism of many spider toxins is through blockage of calcium channels.

A remotely related group of atracotoxins operate by opening sodium channels. Delta atracotoxin produces potentially fatal neurotoxic symptoms in primates by slowing the inactivation of voltage-gated sodium channels. The structure of atracotoxin comprises a core beta region containing a triple-stranded a thumb-like extension protruding from the beta region and a C-terminal helix. The beta region contains a cystine knot motif, a feature seen in other neurotoxic polypeptides and other spider toxins, of the CSTX family.

Spider potassium channel inhibitory toxins is another group of spider toxins. A representative of this group is hanatoxin, a 35 amino acid peptide toxin which was isolated from Chilean rose tarantula (*Grammostola rosea*, syn. *G. spatulata*) venom. It inhibits the drk1 voltage-gated potassium channel by altering the energetics of gating.

Insect Toxin

Insects contain substances called insect toxins, which are synthesized or accumulated from the substrate. According to the different mechanism, these insects can be divided into two categories: Cryptotoxics are insects capable of triggering gastrointestinal reactions or causing the introduction of toxic compounds in the body. They do not contain organ for inoculation of poison as potentially poisonous substances can be located in precise structures or spread throughout the body. Phanerotoxics are insects that have organs for the synthesis, storage and inoculation of poison such as bees or ants. The poison of these insects is usually inactive in the gastrointestinal tract and only active by inoculation. The possible risks associated with this category concern the transit through oral cavity and esophagus. Toxicological tests conducted on insects or on proteins derived from insects, to date, are not yet available. Therefore, for the toxins inherent in the animal, it is necessary to determine which toxins can be eliminated by the insect preparation stages as food and what are the ways to eliminate them.

Instead, the mycotoxins that can be found in insects, derive in most cases from pathogenic fungi present in the substrate, such as Aspergillus spp., Penicillium spp. and Fusarium spp.

References

- Godish, Thad (2001). Indoor environmental quality. Chelsea, Mich: Lewis Publishers. pp. 183–4. ISBN 1-56670-402-2

- Toxin: newworldencyclopedia.org, Retrieved 14 July 2018

- Keller NP, Turner G, Bennett JW (2005). "Fungal secondary metabolism – from biochemistry to genomics". Nat. Rev. Microbiol. 3 (12): 937–47. doi:10.1038/nrmicro1286. PMID 16322742

- Neurotoxin, science: britannica.com, Retrieved 19 May 2018

- Bennett, JW; Klich, M (Jul 2003). "Mycotoxins". Clinical Microbiology Reviews. 16 (3): 497–516. doi:10.1128/CMR.16.3.497-516.2003. PMC 164220. PMID 12857779. Retrieved 31 May 2013

- Venom: newworldencyclopedia.org, Retrieved 10 April 2018

- Bullerman, L., Bianchini, A. (2007). "Stability of mycotoxins during food processing". International Journal of Food Microbiology. 119 (1–2): 140–146. doi:10.1016/j.ijfoodmicro.2007.07.035

- Agricultural-and-biological-sciences, insect-toxin: sciencedirect.com, Retrieved 25 June 2018

- Boonen J, Malysheva S, Taevernier L, Diana Di Mavungu J, De Saeger S, De Spiegeleer B (2012). "Human skin penetration of selected model mycotoxins". Toxicology. 301 (1–3): 21–32. doi:10.1016/j.tox.2012.06.012. PMID 22749975

Chapter 4

Tests for Toxicity

Toxicological testing is conducted in order to determine how damaging a substance is to a living or non-living organism. It is essential for medicine and pesticide testing. The aim of this chapter is to investigate and analyze the varied tests for toxicity, such as in vivo and vitro testing, Draize test, Reinsch test, etc.

In Vivo Testing

In vivo refers to experimentation using a whole, living organism as opposed to a partial or dead organism. Animal studies and clinical trials are two forms of in vivo research. In vivo testing is often employed over in vitro because it is better suited for observing the overall effects of an experiment on a living subject.

While there are many reasons to believe in vivo studies have the potential to offer conclusive insights about the nature of medicine and disease, there is a number of ways that these conclusions can be misleading. For example, a therapy can offer a short-term benefit, but a long-term harm.

Animal Testing

Animal testing is a phrase that most people have heard but are perhaps still unsure of exactly what is involved. Whether it is called animal testing, animal experimentation or animal research, it refers to the experimentation carried out on animals. It is used to assess the safety and effectiveness of everything from medication to cosmetics, as well as understanding how the human body works. While supporters believe it is a necessary practice, those opposed to animal testing believe that it involves the torture and Suffering of Animals.

Understanding Animal Testing

Animal testing is conducted virtually everywhere and its uses are broad. In the UK, standards are quite strict with regards to animal testing and monitoring is similarly rigorous. Animal testing only occurs if there is no other viable alternative to the methods. Animal testing may take place at:

- Universities

- Medical schools

- Pharmaceutical and biotechnology companies

- Military defence establishments

Using Animal Testing

Animal testing is used for countless products and applications. Everything from toiletries to medications have likely been tested on animals at some point prior to their distribution. Some of the products that commonly involve animal testing are:

- Cosmetics

- Drugs

- Food additives

- Supplements

- Household products

- Pesticides

- Industrial chemicals

Animal Testing for Medical Treatments

Virtually every available Medical Treatment today has, to some degree, involved animal testing. The animals themselves may be bred specifically for testing or they may be captured in the wild. There are also commercial establishments that sell animals specifically for use in animal testing facilities. Animals are considered to be similar to humans in terms of assessing safety, which means that there are Strict Requirements for testing on animals with regards to new drugs.

In the UK, for example, a new drug must have been tested on two different species of live mammal. However, those who are opposed to animal testing and view it as an unnecessary infliction of suffering cite that the stress an animal experiences will impact the accuracy of the results, rendering them useless. For now, however, animal testing is required before drugs and some other products are available for consumer use.

Clinical Trials

Clinical trials are scientific studies conducted to find better ways to prevent, screen for, diagnose, or treat disease. These clinical trials may also show which medical approaches work best for certain illnesses or groups of people. Clinical trials produce high-quality data for healthcare decision making.

The purpose of clinical trials is to answer scientific questions. Therefore, these studies follow strict, scientific standards which protect patients and help produce reliable clinical trial results. Clinical trials are one of the final stages of a long and careful research and development process. The process often begins in a laboratory, where scientists first develop and test new ideas.

Clinical trials are divided into different stages, called phases. The earliest phase trials may look at whether a drug is safe or the side effects it causes. A later phase trial aims to test whether a new treatment is better than existing treatments.

There are 3 main phases of clinical trials – phases 1 to 3. But some trials have an earlier stage called phase 0, and there are some phase 4 trials done after a drug has been licensed.

Phase 0 Trials

Phase 1 trials are usually the earliest trials of drugs in people. But your doctor might ask if you would like to join a phase 0 study. These studies aim to find out if a drug behaves in the way researchers expect it to from their laboratory studies.

Phase 0 studies usually only involve a small number of people and they only have a very small dose of a drug. The dose of the drug is too small to treat your cancer, but the types of things researchers are looking for include:

- Whether the drug reaches the cancer
- How the drug behaves in the body
- How cancer cells in the body respond to the drug

You might have extra scans and give extra samples of blood and cancer tissue (biopsies) to help the researchers work out what is happening.

Because the dose of the drug used in phase 0 trials is so small you won't benefit from the drug. But you are also less likely to have side effects.

The main aim of these studies is to speed up the development of promising new drugs. Testing them in very small doses in humans rather than in animals can be more reliable and means scientists get useful information more quickly.

Phase 1 Trials

Phase 1 is sometimes written as phase I. They are usually small trials, recruiting only a few patients. The trial may be open to people with any type of cancer.

When laboratory testing shows that a new treatment might help treat cancer, phase 1 trials are done to find out:

- How much of the drug is safe to give

- What the side effects are

- How the body copes with the drug

- If the treatment shrinks the cancer

Patients are recruited very slowly onto phase 1 trials. So although they don't recruit many patients they can take a long time to complete. The first few patients to take part (called a cohort or group) are given a very small dose of the drug. If all goes well, the next group have a slightly higher dose. The dose is gradually increased with each group. The researchers monitor the effect of the drug until they find the best dose to give. This is called a dose escalation study.

In a phase 1 trial you may have lots of blood tests because the researchers look at how the drug affects you. They also look at how your body copes with, and gets rid of the drug. They record any side effects.

People taking part in phase 1 trials often have advanced cancer. They have usually had all the treatment available to them. They may benefit from the new treatment in the trial but many won't. Phase 1 trials aim to look at doses and side effects. This work has to be done first, before we can test the potential new treatment to see if it works.

Phase 2 Trials

Not all treatments tested in a phase 1 trial make it to a phase 2 trial. Phase 2 is sometimes written as phase II. These trials may be for people who all have the same type of cancer or for people who have different types of cancer.

Phase 2 Trials Aim to Find Out:

- If the new treatment works well enough to test in a larger phase 3 trial

- Which types of cancer the treatment works for

- More about side effects and how to manage them

- More about the best dose to use

Although these treatments have been tested in phase 1 trials, you may still have side effects that the doctors don't know about. Drugs can affect people in different ways.

Phase 2 trials are often larger than phase 1. There may be up to 100 or so people taking part. Sometimes in a phase 2 trial, a new treatment is compared with another treatment already in use, or with a dummy drug (placebo). If the results of phase 2 trials show that a new treatment may be as good as existing treatment, or better, it then moves into phase 3.

Some phase 2 trials are randomized. This means the researchers put the people taking part into treatment groups at random.

Phase 3 Trials

These trials compare new treatments with the best currently available treatment (the standard treatment). Phase 3 is sometimes written as phase III. These trials may compare:

- A completely new treatment with the standard treatment
- Different doses or ways of giving a standard treatment
- A new way of giving radiotherapy with the standard way

Phase 3 trials usually involve many more patients than phase 1 or 2. This is because differences in success rates may be small. So, the trial needs many patients to be able to show the difference.

One example could be that 6 out of 100 more people (6%) get a remission Open a glossary item with a new treatment compared to standard treatment. If there were 50 people in the new treatment group and 50 people in the standard treatment group, there may be 3 more people in remission in the new treatment group. The 2 groups would not look that different. But if the researchers gave each treatment to 5,000 people, there could be 300 more remissions in the new treatment group.

Sometimes phase 3 trials involve thousands of patients in many different hospitals and even different countries. Most phase 3 trials are randomized. This means the researchers put the people taking part into treatment groups at random.

Phase 4 Trials

Phase 4 trials are done after a drug has been shown to work and has been granted a license. Phase 4 is sometimes written as phase IV. The main reasons for running phase 4 trials are to find out:

- More about the side effects and safety of the drug
- What the long term risks and benefits are
- How well the drug works when it's used more widely

Trials Covering more than One Phase

Most trials are just one phase. But some trials cover more than one phase. For example, the same trial can include both phase 1 and phase 2. The aim of phase 1 might be to work out the highest safe dose of a new drug. And the aim of phase 2 might be to see how well that dose works. So you may see trials written as phase 1/2 or phase 2/3.

Multi-arm Multi-stage (MAMS) Trials

Many trials look at just one new treatment. But some trials compare several treatments. The people taking part are still put into groups at random but there may be 3 or more treatment groups. This is a multi-arm trial.

Some trials are designed so that they stop recruiting into a particular group if early results show that a treatment isn't working as well as the others, or is causing more side effects. This is called a multi-arm multi-stage (MAMS) trial. A MAMS trial may also be able to add new groups to look at more treatments.

Administration

Clinical trials designed by a local investigator, and (in the US) federally funded clinical trials, are almost always administered by the researcher who designed the study and applied for the grant. Small-scale device studies may be administered by the sponsoring company. Clinical trials of new drugs are usually administered by a contract research organization (CRO) hired by the sponsoring company. The sponsor provides the drug and medical oversight. A CRO is contracted to perform all the administrative work on a clinical trial. For phases 2, 3 and 4, the CRO recruits participating researchers, trains them, provides them with supplies, coordinates study administration and data collection, sets up meetings, monitors the sites for compliance with the clinical protocol, and ensures the sponsor receives data from every site. Specialist site management organizations can also be hired to coordinate with the CRO to ensure rapid IRB/IEC approval and faster site initiation and patient recruitment. Phase 1 clinical trials of new medicines are often conducted in a specialist clinical trial clinic, with dedicated pharmacologists, where the subjects can be observed by full-time staff. These clinics are often run by a CRO which specializes in these studies.

At a participating site, one or more research assistants (often nurses) do most of the work in conducting the clinical trial. The research assistant's job can include some or all of the following: providing the local institutional review board (IRB) with the documentation necessary to obtain its permission to conduct the study, assisting with study start-up, identifying eligible patients, obtaining consent from them or their families, administering study treatment(s), collecting and statistically analyzing data, maintaining and updating data files during followup, and communicating with the IRB, as well as the sponsor and CRO.

Marketing

Janet Yang uses the Interactional Justice Model to test the effects of willingness to talk with a doctor and clinical trial enrollment. Results found that potential clinical trial candidates were less likely to enroll in clinical trials if the patient is more willing to talk with their doctor. The reasoning behind this discovery may be patients are happy with their current care. Another reason for the negative relationship between perceived fairness and clinical trial enrollment is the lack of independence from the care provider. Results found that there is a positive relationship between a lack of willingness to talk with their doctor and clinical trial enrollment. Lack of willingness to talk about clinical trials with current care providers may be due to patients' independence from the doctor. Patients who are less likely to talk about clinical trials are more willing to use other sources of information to gain a better insight of alternative treatments. Clinical trial enrollment should be motivated to utilize websites and television advertising to inform the public about clinical trial enrollment.

Information Technology

The last decade has seen a proliferation of information technology use in the planning and conduct of clinical trials. Clinical trial management systems are often used by research sponsors or CROs to help plan and manage the operational aspects of a clinical trial, particularly with respect to investigational sites. Advanced analytics for identifying researchers and research sites with expertise in a given area utilize public and private information about ongoing research. Web-based electronic data capture (EDC) and clinical data management systems are used in a majority of clinical trials to collect case report data from sites, manage its quality and prepare it for analysis. Interactive voice response systems are used by sites to register the enrollment of patients using a phone and to allocate patients to a particular treatment arm (although phones are being increasingly replaced with web-based (IWRS) tools which are sometimes part of the EDC system). While patient-reported outcome were often paper based in the past, measurements are increasingly being collected using web portals or hand-held ePRO (or eDiary) devices, sometimes wireless. Statistical software is used to analyze the collected data and prepare them for regulatory submission. Access to many of these applications are increasingly aggregated in web-based clinical trial portals. In 2011, the FDA approved a phase 1 trial that used telemonitoring, also known as remote patient monitoring, to collect biometric data in patients' homes and transmit it electronically to the trial database. This technology provides many more data points and is far more convenient for patients, because they have fewer visits to trial sites.

Ethical Aspects

Clinical trials are closely supervised by appropriate regulatory authorities. All studies involving a medical or therapeutic intervention on patients must be approved by a supervising ethics committee before permission is granted to run the trial. The local ethics committee has discretion on how it will supervise noninterventional studies (observational studies or those using already collected data). In the US, this body is called the Institutional Review Board (IRB); in the EU, they are called Ethics committees. Most IRBs are located at the local investigator's hospital or institution, but some sponsors allow the use of a central (independent/for profit) IRB for investigators who work at smaller institutions.

To be ethical, researchers must obtain the full and informed consent of participating human subjects. (One of the IRB's main functions is to ensure potential patients are adequately informed about the clinical trial.) If the patient is unable to consent for him/ herself, researchers can seek consent from the patient's legally authorized representative. In California, the state has prioritized the individuals who can serve as the legally authorized representative.

In some US locations, the local IRB must certify researchers and their staff before they can conduct clinical trials. They must understand the federal patient privacy (HIPAA) law and good clinical practice. The International Conference of Harmonization Guidelines for Good Clinical Practice is a set of standards used internationally for the conduct of clinical trials. The guidelines aim to ensure the "rights, safety and well being of trial subjects are protected".

The notion of informed consent of participating human subjects exists in many countries all over the world, but its precise definition may still vary.

Informed consent is clearly a 'necessary' condition for ethical conduct but does not 'ensure' ethical conduct. In compassionate use trials the latter becomes a particularly difficult problem. The final objective is to serve the community of patients or future patients in a best-possible and most responsible way. However, it may be hard to turn this objective into a well-defined, quantified, objective function. In some cases this can be done, however, for instance, for questions of when to stop sequential treatments, and then quantified methods may play an important role.

Additional ethical concerns are present when conducting clinical trials on children (pediatrics), and in emergency or epidemic situations.

Conflicts of Interest and Unfavorable Studies

In response to specific cases in which unfavorable data from pharmaceutical company-sponsored research were not published, the Pharmaceutical Research and Manufacturers of America published new guidelines urging companies to report all findings and limit the financial involvement in drug companies by researchers. The US Congress signed into law a bill which requires phase II and phase III clinical trials to be registered by the sponsor on the clinicaltrials.gov website compiled by the National Institutes of Health.

Drug researchers not directly employed by pharmaceutical companies often seek grants from manufacturers, and manufacturers often look to academic researchers to conduct studies within networks of universities and their hospitals, e.g., for translational cancer research. Similarly, competition for tenured academic positions, government grants and prestige create conflicts of interest among academic scientists. According to one study, approximately 75% of articles retracted for misconduct-related reasons have no declared industry financial support. Seeding trials are particularly controversial.

In the United States, all clinical trials submitted to the FDA as part of a drug approval process are independently assessed by clinical experts within the Food and Drug Administration, including inspections of primary data collection at selected clinical trial sites.

Safety

Responsibility for the safety of the subjects in a clinical trial is shared between the sponsor, the local site investigators (if different from the sponsor), the various IRBs that supervise the study, and (in some cases, if the study involves a marketable drug or device), the regulatory agency for the country where the drug or device will be sold.

For safety reasons, many clinical trials of drugs are designed to exclude women of childbearing age, pregnant women, or women who become pregnant during the study. In some cases, the male partners of these women are also excluded or required to take birth control measures.

Sponsor

Throughout the clinical trial, the sponsor is responsible for accurately informing the local site investigators of the true historical safety record of the drug, device or other medical treatments to be tested, and of any potential interactions of the study treatment(s) with already approved treatments. This allows the local investigators to make an informed judgment on whether to participate in the study or not. The sponsor is also responsible for monitoring the results of the study as they come in from the various sites as the trial proceeds. In larger clinical trials, a sponsor will use the services of a data monitoring committee (DMC, known in the US as a data safety monitoring board). This independent group of clinicians and statisticians meets periodically to review the unblinded data the sponsor has received so far. The DMC has the power to recommend termination of the study based on their review, for example if the study treatment is causing more deaths than the standard treatment, or seems to be causing unexpected and study-related serious adverse events. The sponsor is responsible for collecting adverse event reports from all site investigators in the study, and for informing all the investigators of the sponsor's judgment as to whether these adverse events were related or not related to the study treatment.

The sponsor and the local site investigators are jointly responsible for writing a site-specific informed consent that accurately informs the potential subjects of the true risks and potential benefits of participating in the study, while at the same time presenting the material as briefly as possible and in ordinary language. FDA regulations state that participating in clinical trials is voluntary, with the subject having the right not to participate or to end participation at any time.

Local Site Investigators

The ethical principle of *primum non nocere* ("first, do no harm") guides the trial, and if an investigator believes the study treatment may be harming subjects in the study, the investigator can stop participating at any time. On the other hand, investigators often have a financial interest in recruiting subjects, and could act unethically to obtain and maintain their participation.

The local investigators are responsible for conducting the study according to the study protocol, and supervising the study staff throughout the duration of the study. The local investigator or his/her study staff are also responsible for ensuring the potential subjects in the study understand the risks and potential benefits of participating in the study. In other words, they (or their legally authorized representatives) must give truly informed consent.

Local investigators are responsible for reviewing all adverse event reports sent by the sponsor. These adverse event reports contain the opinion of both the investigator at the site where the adverse event occurred, and the sponsor, regarding the relationship of the adverse event to the study treatments. Local investigators also are responsible for making an independent judgment of these reports, and promptly informing the local IRB of all serious and study treatment-related adverse events.

When a local investigator is the sponsor, there may not be formal adverse event reports, but study staff at all locations are responsible for informing the coordinating investigator of anything unexpected. The local investigator is responsible for being truthful to the local IRB in all communications relating to the study.

Institutional Review Boards (IRBs)

Approval by an Institutional Review Board (IRB), or ethics board, is necessary before all but the most informal research can begin. In commercial clinical trials, the study protocol is not approved by an IRB before the sponsor recruits sites to conduct the trial. However, the study protocol and procedures have been tailored to fit generic IRB submission requirements. In this case, and where there is no independent sponsor, each local site investigator submits the study protocol, the consent(s), the data collection forms, and supporting documentation to the local IRB. Universities and most hospitals have in-house IRBs. Other researchers (such as in walk-in clinics) use independent IRBs.

The IRB scrutinizes the study for both medical safety and protection of the patients involved in the study, before it allows the researcher to begin the study. It may require changes in study procedures or in the explanations given to the patient. A required yearly "continuing review" report from the investigator updates the IRB on the progress of the study and any new safety information related to the study.

Regulatory Agencies

In the US, the FDA can audit the files of local site investigators after they have finished participating in a study, to see if they were correctly following study procedures. This audit may be random, or for cause (because the investigator is suspected of fraudulent data). Avoiding an audit is an incentive for investigators to follow study procedures.

Alternatively, many American pharmaceutical companies have moved some clinical trials overseas. Benefits of conducting trials abroad include lower costs (in some countries) and the ability to run larger trials in shorter timeframes, whereas a potential disadvantage exists in lower-quality trial management. Different countries have different regulatory requirements and enforcement abilities. An estimated 40% of all clinical trials now take place in Asia, Eastern Europe, and Central and South America. "There is no compulsory registration system for clinical trials in these countries and many do not follow European directives in their operations", says Jacob Sijtsma of the Netherlands-based WEMOS, an advocacy health organisation tracking clinical trials in developing countries.

Beginning in the 1980s, harmonization of clinical trial protocols was shown as feasible across countries of the European Union. At the same time, coordination between Europe, Japan and the United States led to a joint regulatory-industry initiative on international harmonization named after 1990 as the International Conference on Harmonisation of Technical Requirements for Registration of Pharmaceuticals for Human Use (ICH) Currently, most clinical trial programs follow ICH guidelines, aimed at "ensuring that good quality, safe and effective medicines are developed and registered in the most efficient and cost-effective manner. These activities are pursued in the interest of the consumer and public health, to prevent unnecessary duplication of clinical trials in humans and to minimize the use of animal testing without compromising the regulatory obligations of safety and effectiveness."

Aggregation of Safety Data During Clinical Development

Aggregating safety data across clinical trials during drug development is important because trials are generally designed to focus on determining how well the drug works. The safety data collected and aggregated across multiple trials as the drug is developed allows the sponsor, investigators and regulatory agencies to monitor the aggregate safety profile of experimental medicines as they're developed. The value of assessing aggregate safety data is:

a) Decisions based on aggregate safety assessment during development of the medicine can be made throughout the medicine's development;

b) It sets up the sponsor and regulators well for assessing the medicine's safety after the drug is approved.

Sponsor

The cost of a study depends on many factors, especially the number of sites conducting the study, the number of patients involved, and whether the study treatment is already approved for medical use.

The expenses incurred by a pharmaceutical company in administering a phase 3 or 4 clinical trial may include, among others:

- Production of the drug(s) or device(s) being evaluated

- Staff salaries for the designers and administrators of the trial

- Payments to the contract research organization, the site management organization (if used) and any outside consultants

- Payments to local researchers and their staff for their time and effort in recruiting test subjects and collecting data for the sponsor

- The cost of study materials and the charges incurred to ship them

- Communication with the local researchers, including on-site monitoring by the CRO before and (in some cases) multiple times during the study

- One or more investigator training meetings

- Expense incurred by the local researchers, such as pharmacy fees, IRB fees and postage

- Any payments to subjects enrolled in the trial

- The expense of treating a test subject who develops a medical condition caused by the study drug

These expenses are incurred over several years.

In the USA, sponsors may receive a 50 percent tax credit for clinical trials conducted on drugs being developed for the treatment of orphan diseases. National health agencies, such as the US National Institutes of Health, offer grants to investigators who design clinical trials that attempt to answer research questions of interest to the agency. In these cases, the investigator who writes the grant and administers the study acts as the sponsor, and coordinates data collection from any other sites. These other sites may or may not be paid for participating in the study, depending on the amount of the grant and the amount of effort expected from them. Using internet resources can, in some cases, reduce the economic burden.

Investigators

Investigators are often compensated for their work in clinical trials. These amounts can be small, just covering a partial salary for research assistants and the cost of any supplies (usually the case with national health agency studies), or be substantial and include 'overhead' that allows the investigator to pay the research staff during times between clinical trials.

Subjects

Participants in phase 1 drug trials do not gain any direct health benefit from taking part. They are generally paid a fee for their time, with payments regulated and not related to any risk involved. In later phase trials, subjects may not be paid to ensure their motivation for participating with potential for a health benefit or contributing to medical knowledge. Small payments may be made for study-related expenses such as travel or as compensation for their time in providing follow-up information about their health after the trial treatment ends.

Participant Recruitment and Participation

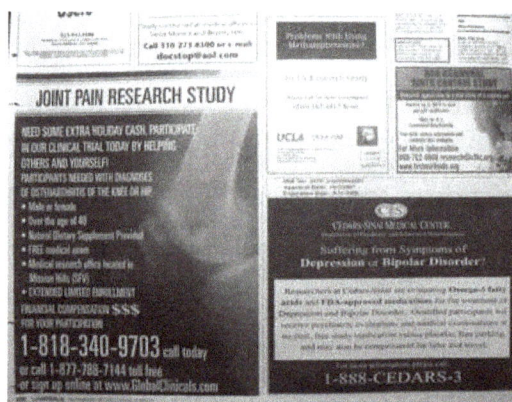

Newspaper advertisements seeking patients and healthy volunteers to participate in clinical trials

Phase 0 and phase 1 drug trials seek healthy volunteers. Most other clinical trials seek patients who have a specific disease or medical condition. The diversity observed in society should be reflected in clinical trials through the appropriate inclusion of ethnic minority populations. Patient recruitment or participant recruitment plays a significant role in the activities and responsibilities of sites conducting clinical trials.

All volunteers being considered for a trial are required to undertake a medical screening. Requirements differ according to the trial needs, but typically volunteers would be screened in a medical laboratory for:

- Measurement of the electrical activity of the heart (ECG)
- Measurement of blood pressure, heart rate and body temperature
- Blood sampling
- Urine sampling

- Weight and height measurement

- Drug abuse testing

- Pregnancy testing

Locating Trials

Depending on the kind of participants required, sponsors of clinical trials, or contract research organizations working on their behalf, try to find sites with qualified personnel as well as access to patients who could participate in the trial. Working with those sites, they may use various recruitment strategies, including patient databases, newspaper and radio advertisements, flyers, posters in places the patients might go (such as doctor's offices), and personal recruitment of patients by investigators.

Volunteers with specific conditions or diseases have additional online resources to help them locate clinical trials. For example, the Fox Trial Finder connects Parkinson's disease trials around the world to volunteers who have a specific set of criteria such as location, age, and symptoms. Other disease-specific services exist for volunteers to find trials related to their condition. Volunteers may search directly on ClinicalTrials.gov to locate trials using a registry run by the U.S. National Institutes of Health and National Library of Medicine.

In Vitro Testing

A test performed *in vitro* means that it is done outside of a living organism and it usually involves isolated tissues, organs or cells.

You can use *in vitro* data to fully or partly fulfill information requirements that would otherwise need data to be generated with tests on living organisms (*in vivo* tests).

In vitro methods are divided to those that meet internationally agreed validation criteria and to those that do not. For your REACH registration, preferably use methods that are sufficiently well-developed according to internationally agreed test development criteria (e.g. the European Centre for the Validation of Alternative Methods (ECVAM) pre-validation criteria).

The changes to the REACH Annexes VII and VIII in 2016 make *in vitro* test methods the default for certain toxicological properties. In most cases, no new *in vivo* studies are necessary for registrations below 100 tons for the classification or risk assessment of your substance.

Even if you have used an *in vitro* method that has not been internationally validated, you still need to submit your study results in your registration dossier as part of gathering all available information. You may also use them in a weight of evidence approach or to support the grouping of your substances.

Tips

- Present a clear and detailed description of the results, the test conditions and the interpretation of the results. This is important if the study is used as a key study or as part of a weight of evidence approach.

- Communicate clearly the limitations of the method.

- Use a (preferably) validated method. Make sure that the results are adequate for classification and labeling and risk assessment and that you provide adequate and reliable documentation.

- Your *in vitro* method needs to deliver reliable information that is comparable to that from the standard test. If not, you may then need *in vivo* testing.

In Vitro to in Vivo Extrapolation

In vitro to in vivo extrapolation (IVIVE) refers to the qualitative or quantitative transposition of experimental results or observations made in vitro to predicts phenomena in vivo, biological organisms.

The problem of transposing in vitro results is particularly acute in areas such as toxicology where animal experiments are being phased out and are increasingly being replaced by alternative tests.

Results obtained from *in vitro* experiments cannot often be directly applied to predict biological responses of organisms to chemical exposure *in vivo*. Therefore, it is extremely important to build a consistent and reliable *in vitro* to *in vivo* extrapolation method.

Two solutions are now commonly accepted:

- Increasing the complexity of *in vitro* systems where multiple cells can interact with each other in order recapitulate cell-cell interactions present in tissues (as in "human on chip" systems).

- Using mathematical modeling to numerically simulate the behavior of a complex system, whereby *in vitro* data provides the parameter values for developing a model.

The two approaches can be applied simultaneously allowing in vitro systems to provide adequate data for the development of mathematical models. To comply with push for the development of alternative testing methods, increasingly sophisticated in vitro experiments are now collecting numerous, complex, and challenging data that can be integrated into mathematical models.

Pharmacology

IVIVE in pharmacology can be used to assess pharmacokinetics (PK) or pharmacodynamics (PD).

Since biological perturbation depends on concentration of the toxicant as well as exposure duration of a candidate drug (parent molecule or metabolites) at that target site, in vivo tissue and organ effects can either be completely different or similar to those observed *in vitro*. Therefore, extrapolating adverse effects observed in vitro is incorporated into a quantitative model of in vivo PK model. It is generally accepted that physiologically based PK (PBPK) models, including absorption, distribution, metabolism, and excretion of any given chemical are central to *in vitro - in vivo* extrapolations.

In the case of early effects or those without inter-cellular communications, it is assumed that the same cellular exposure concentration cause the same effects, both experimentally and quantitatively, *in vitro* and *in vivo*. In these conditions, it is enough to:

(1) Develop a simple pharmacodynamics model of the dose–response relationship observed *in vitro*,

(2) Transpose it without changes to predict *in vivo* effects.

However, cells in cultures do not mimic perfectly cells in a complete organism. To solve that extrapolation problem, more statistical models with mechanistic information are needed, or we can rely on mechanistic systems of biology models of the cell response. Those models are characterized by a hierarchical structure, such as molecular pathways, organ function, whole-cell response, cell-to- cell communications, tissue response and inter-tissue communications.

DEBtox

DEBtox is a user-friendly software package designed to analyze the results of the standard set of aquatic toxicity tests on the basis of the Dynamic Energy Budget theory:

- Acute and chronic tests on survival/immobility
- The (37d) fish growth test
- The (21d) reproduction test with *Daphnia*
- The (4d) alga growth inhibition test

The tests conform to OECD (Organization for Economic Co-operation and Development) and ISO (International Organization for Standardization) standards.

Reinsch Test

The Reinsch test is an initial indicator to detect the presence of one or more of the following heavy metals in a biological sample, and is often used by toxicologists where poisoning by such metals is suspected. The method which is sensitive to Antimony, Arsenic, Bismuth, Selenium, Thallium and Mercury was discovered by Hugo Reinsch in 1841.

Process

- Dissolve suspect body fluid or tissue in a hydrochloric acid solution.

- Insert a copper strip into the solution.

- The appearance of a silvery coating on the copper may indicate Mercury. A dark coating indicative of the presence of one of the other metals.

- Confirm finding using absorption or emission spectroscopy, X-ray diffraction, or other analytical technique suitable for inorganic analysis.

- A scientific application of the Reinsch Test was presented in 2010 by chemists of Technische Universität München, Lehrstuhl für Radiochemie (Institute for Radiochemistry, Technical University Munich) and ITU (Institute for Transuranium Elements, Karlsruhe): in the course of the radiochemical purification of Se for the determination of its half-life, reductive deposition of selenium on metallic copper was the first step to extract Se from high active raffinate (= PUREX raffinate) in a hot cell.

Draize Test

In 1944, the Draize test was invented as a way to measure skin and eye irritancy of chemicals and other products. It is an important test for measuring toxicity. Specifically, this test involves dropping concentrated amounts of a test substance into an animal's eye (while their lids are clipped open) or placing a chemical onto an area where the animal's skin has been shaved. The resulting irritation, which may include ulceration, inflamed/bleeding skin, swollen eyes, and blindness, is subsequently measured on a numerical scale.

This test has been heavily criticized for shortfalls in predictability, reproducibility, and subjectivity. Although it is performed on a relatively small number of animals, the

Draize test is considerably one of the cruelest test methods that individual animals are forced to endure. Victims of this test may be immobilized for up to 14 days, sometimes without any pain medication.

Numerous in vitro tests for skin and eye irritation have been developed, validated, and internationally recognized by agencies including ICCVAM, ECCVAM, and OECD. The tests include methods that are likely to be more predictive of human responses (i.e. synthetic skin) and are undeniably more efficient and humane.

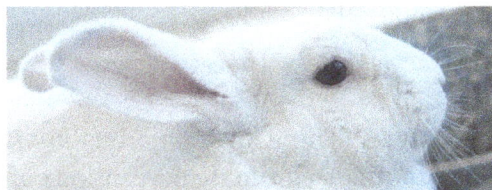

Draize Predictability Argument

The Draize test is a poor predictor of human response because of structural differences between human and rabbit eyes and due to the subjective analysis of data. To illustrate this point, here are some facts:

- A rabbit's eye is anatomically and physiologically different than a human's eye. For example, rabbits produce fewer tears than humans do, so their eyes cannot easily flush out test chemicals. A rabbit's cornea is substantially thinner and more easily damaged than a human cornea and covers a greater surface area (25%) than that of a human eye (7%).

- A 2004 study by the U.S. Scientific Advisory Committee on Alternative Toxicological Methods analyzed the modern Draize skin test, and found that tests were 10.3%-38.7% likely to misidentify a serious irritant as a mild irritant.

- The Draize method of assessing eye and skin damage is based on the investigator's subjective interpretation of damage, and results may vary significantly between individual investigators and laboratories. Therefore, the data may be unreliable.

Impact

Since 2005, the FDA has stated that the Draize test data was no longer needed for primary skin and eye irritancy. Despite no longer required by the FDA, the Draize test is still evidentially being used. From 2011-2014 the FDA approved 137 new compounds, 24% of which were tested for skin irritancy and 22% for eye irritancy. The Draize test was used in 94% of all skin irritation and 60% of all eye irritation testing.

Following a protest campaign in the 1980s, consumer awareness drove many companies to abandon the test altogether. The rise of cruelty-free cosmetics/personal care companies has also contributed to reduction of Draize usage. However, there are still hundreds of other chemicals that fall outside the scope of cosmetics/personal care, including: dishwashing liquid, cleaning agents, pesticides, and other product ingredients.

Although the total number of animals used in the Draize test is relatively small, the level of suffering per animal is substantial. It has been described as "the most cruel and infamous experiments on rabbits known to mankind." Test subjects are immobilized and therefore restrained from pawing at their eyes or skin to relieve discomfort. Furthermore, anesthesia is not required.

Feasibility

- Availability of Alternatives: There are numerous in vitro and ex vivo replacements which could potentially eliminate perceptions of necessity for the Draize test. These alternatives have been scientifically validated and accepted to replace animal tests for eye and skin irritation. Examples include:

  o In vitro: In 2013, the development of a new in vitro eye testing method called the Human Corneal Epithelium model was announced which uses actual human cornea cells and is therefore expected to produce more predictive results.

  o Ex vivo: alternatives include ICE (Isolated Chicken Eye) test and BCOP (Bovine Corneal Opacity and Permeability) test.

- Current Use: Although the FDA stated that the Draize test was no longer "required," companies have continued to utilize it at their discretion. Considering the availability of alternative test methods, possible reasons for their continued usage include: lack of information and accessibility, industrial pressures, and subjective resistance to change.

- Reduction/Replacement Trend: The availability of the Draize test has been substantially reduced in other countries through legislation that prohibits animal testing for cosmetics. For example, India passed a law that specifically mandates replacement of the Draize test with alternative test methods.

Early Life Stage Test

An early life stage (ELS) test is a chronic toxicity test using sensitive early life stages like embryos or larvae to predict the effects of toxicants on organisms. ELS tests were developed to be quicker and more cost-efficient than full life-cycle tests, taking on average 1–5 months to complete compared to 6–12 months for a life-cycle test. They are commonly used in aquatic toxicology, particularly with fish. Growth and survival are the typically measured endpoints, for which a Maximum Acceptable Toxicant

Concentration (MATC) can be estimated. ELS tests allow for the testing of fish species that otherwise could not be studied due to length of life, spawning requirements, or size. ELS tests are used as part of environmental risk assessments by regulatory agencies including the U.S. Environmental Protection Agency (EPA) and Environment Canada, as well as the Organisation for Economic Co-operation and Development (OECD).

Development

ELS tests were adapted from full life-cycle toxicity tests, chronic tests that expose an organism to a contaminant for its entire life-cycle. These are widely considered to be the best tests for estimating long-term "safe" concentrations of toxicants in aquatic organisms. The first full life-cycle tests on fish were developed for the fathead minnow (*Pimephales promelas*), and later for bluegill (*Lepomis macrochirus*), brook trout (*Salvelinus fontinalis*), flagfish (*Jordanella floridae*), and sheepshead minnow (*Cyprinodon variegatus*). While useful, full life-cycle tests require a high number of test organisms and extensive exposure time in the lab, especially for vertebrates. Typically, life-cycle tests take 6–12 months for fathead minnow and 30 months for brook trout.

Following the passage of the Toxic Substances Control Act (TSCA) in the United States in 1976, there was an increased need for quicker, more efficient vertebrate toxicity tests. The EPA was now required to assess the environmental effects of new chemicals before they could be commercially produced. Less costly and time-intensive tests were needed to evaluate a multitude of new chemicals. Researchers began developing toxicity tests that focused on early life stages, since these have been shown to be more sensitive to environmental stressors than later life stages. Many critical events occur in a short period of time in the early stages of development. If a stressor disrupts developmental events (including their timing), it could result in adverse effects that reduce the organism's chances of survival. Meta-analysis has found that early life-cycle portions of full life-cycle tests usually estimate an MATC within a factor of 2 of full life-cycle estimates in saltwater and freshwater fish. In 83% of 72 tests, the ELS portion resulted in the same MATC as the full life-cycle estimate, and the remaining 17% were within a factor of 2.

Limitations

There remain some limitations with early life stage toxicity testing. Although ELS tests are quicker and more cost-efficient than full life cycle tests, they remain resource- and time-intensive. Fish early life stage (FELS) tests require hundreds of fish and 1 to 5 months to complete. Other issues include the lack of mechanistic information, differing sensitivities between species, and insensitivity to parental exposure. ELS tests don't provide information on the toxicant's mechanism of action. Sensitivity to specific toxicants varies with species, so the most sensitive or most important species should be tested in each case. ELS tests appear to be insensitive to parental exposure, and MATCs are generally the same for embryos of both exposed and unexposed parents. This could be due to the mode of action of the toxicant or the variability and insensitivity of ELS test design. Additionally, growth response has been found to be an insensitive endpoint in ELS tests with fish, having little bearing on the estimation of an MATC. Growth response could be omitted to reduce the duration and cost of screening tests.

Methodology

In a typical early life stage toxicity test, a flow-through dilutor system administers different concentrations of a toxicant to different test chambers. At least five different concentrations of a toxicant are tested, plus controls, with at least two exposure chambers for each treatment. The length of the exposure depends on the test species. For example, fathead minnow tests are 1–2 months long, while brook trout tests are around 5 months long. Growth and survival are the typical endpoints, for which an MATC can be found.

Standard methods for ELS tests have been established by the OECD, ASTM International, the EPA, and Environment Canada.

Regulatory Uses

- The FELS guideline of the OECD Guidelines for the Testing of Chemicals is the primary test used internationally to estimate chronic fish toxicity.

- FELS tests are part of a suite of sublethal toxicity tests for effluent used by Environment Canada in environmental effects monitoring.

- A FELS test is required or recommended by the US EPA for testing and monitoring chemicals released into aquatic systems.

Current Developments

An extended ELS test has been examined as a potential surrogate for a fish full life-cycle test to detect weak environmental estrogens. Endocrine active chemicals (EACs) are ubiquitous in the environment, prompting the need for better screening assays to predict their effects, especially in aquatic species. Slightly longer ELS tests could be used instead of full life-cycle tests, taking into account sensitive windows of exposure like sexual differentiation and early gonadal development. Extended ELS tests have proven successful in detecting the effects of weak estrogens in fathead minnows.

Additionally, adverse outcome pathways (AOPs) are being used to develop an alternative to FELS testing. Industry and regulatory agencies are increasingly interested in an animal-free, cost-efficient surrogate. Researchers are developing FELS-related AOPs to create a high-throughput, less costly screening strategy for toxicants that takes the mechanism of action into account.

Marsh Test

Marsh test is a method for the detection of arsenic. It is so sensitive that it can be used to detect minute amounts of arsenic in foods (the residue of fruit spray) or in stomach contents. The sample is placed in a flask with arsenic-free zinc and sulfuric acid. Arsine gas (also hydrogen) forms and is led through a drying tube to a hard glass tube in which it is heated. The arsenic is deposited as a mirror just beyond the heated area and on any cold surface held in the burning gas emanating from the jet. Antimony gives a similar test, but the deposit is insoluble in sodium hypochlorite, whereas arsenic will dissolve. The test was named for its inventor, the English chemist James Marsh.

Precursor Methods

The first breakthrough in the detection of arsenic poisoning was in 1775 when Carl Wilhelm Scheele discovered a way to change arsenic trioxide to garlic-smelling arsine gas (AsH_3), by treating it with nitric acid (HNO_3) and combining it with zinc.

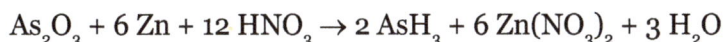

$$As_2O_3 + 6\ Zn + 12\ HNO_3 \rightarrow 2\ AsH_3 + 6\ Zn(NO_3)_2 + 3\ H_2O$$

In 1787, German physician Johann Metzger discovered that if arsenic trioxide were heated in the presence of carbon, the arsenic would sublime. This is the reduction of As_2O_3 by carbon:

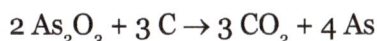

$$2\ As_2O_3 + 3\ C \rightarrow 3\ CO_2 + 4\ As$$

In 1806, Valentin Rose took the stomach of a victim suspected of being poisoned and treated it with potassium carbonate (K_2CO_3), calcium oxide (CaO) and nitric acid. Any arsenic present would appear as arsenic trioxide and then could be subjected to Metzger's test.

However, the most common test (and used even today in water test kits) was discovered by Samuel Hahnemann. It would involve combining a sample fluid with hydrogen sulfide (H_2S) in the presence of hydrochloric acid (HCl). A yellow precipitate, arsenic trisulfide (As_2S_3) would be formed if arsenic was present.

Circumstances and Methodology

Even so, these tests have proven not to be sensitive enough. In 1832, a certain John Bodle was brought to trial for poisoning his grandfather by putting arsenic in his coffee. James Marsh, a chemist working at the Royal Arsenal in Woolwich, was called by the prosecution to try to detect its presence. He performed the standard test by passing hydrogen sulfide through the suspect fluid. While Marsh was able to detect arsenic, the yellow precipitate did not keep very well, and, by the time it was presented to the jury, it had deteriorated. The jury was not convinced, and John Bodle was acquitted.

Angered and frustrated by this, especially when John Bodle confessed later that he indeed killed his grandfather, Marsh decided to devise a better test to demonstrate the presence of arsenic. Taking Scheele's work as a basis, he constructed a simple glass apparatus capable of not only detecting minute traces of arsenic but also measuring its quantity. Adding a sample of tissue or body fluid to a glass vessel with zinc and acid would produce arsine gas if arsenic was present, in addition to the hydrogen that would be produced regardless by the zinc reacting with the acid. Igniting this gas mixture would oxidize any arsine present into arsenic and water vapor. This would cause a cold ceramic bowl held in the jet of the flame to be stained with a silvery-black deposit of arsenic, physically similar to the result of Metzger's reaction. The intensity of the stain could then be compared to films produced using known amounts of arsenic. Not only could minute amounts of arsenic be detected (as little as 0.02 mg), the test was very specific for arsenic. Although antimony (Sb) could give a false-positive test by forming stibine (SbH_3) gas which decomposes on heating to form a similar black deposit, it would not dissolve in a solution of sodium hypochlorite (NaOCl), while arsenic would.

Specific Reactions Involved

The Marsh test treats the sample with sulfuric acid and arsenic-free zinc. Even if there are minute amounts of arsenic present, the zinc reduces the trivalent arsenic (As^{3+}). Here are the two half-reactions:

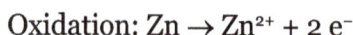

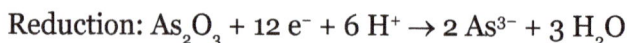

Oxidation: $Zn \rightarrow Zn^{2+} + 2\,e^-$

Reduction: $As_2O_3 + 12\,e^- + 6\,H^+ \rightarrow 2\,As^{3-} + 3\,H_2O$

Overall, we have this reaction:

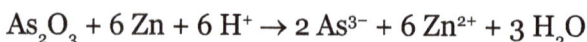

$As_2O_3 + 6\,Zn + 6\,H^+ \rightarrow 2\,As^{3-} + 6\,Zn^{2+} + 3\,H_2O$

In an acidic medium, As^{3-} is protonated to form arsine gas (AsH_3), so adding sulphuric acid (H_2SO_4) to each side of the equation we get:

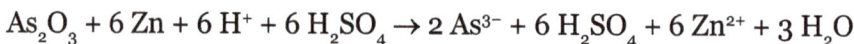

$As_2O_3 + 6\,Zn + 6\,H^+ + 6\,H_2SO_4 \rightarrow 2\,As^{3-} + 6\,H_2SO_4 + 6\,Zn^{2+} + 3\,H_2O$

As the As^{3-} combines with the H^+ to form arsine:

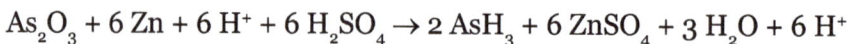

$As_2O_3 + 6\,Zn + 6\,H^+ + 6\,H_2SO_4 \rightarrow 2\,AsH_3 + 6\,ZnSO_4 + 3\,H_2O + 6\,H^+$

By eliminating the common ions:

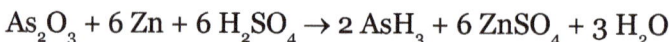

$As_2O_3 + 6\,Zn + 6\,H_2SO_4 \rightarrow 2\,AsH_3 + 6\,ZnSO_4 + 3\,H_2O$

First Notable Application

Although the Marsh test was efficacious, its first publicly documented use—in fact, the first time evidence from forensic toxicology was ever introduced—was in Tulle, France in 1840 with the celebrated LaFarge poisoning case. Charles LaFarge, a foundry owner, was suspected of being poisoned with arsenic by his wife Marie. The circumstantial evidence was great: it was shown that she bought arsenic trioxide from a local chemist, supposedly to kill rats which infested their home. In addition, their maid swore that she had mixed a white powder into his drink. Although the food was found to be positive for the poison using the old methods as well as the Marsh test, when the husband's body was exhumed and tested, the chemists assigned to the case were not able to detect arsenic. Mathieu Orfila, the renowned toxicologist retained by the defense and an acknowledged authority of the Marsh test, examined the results. He performed the test again, and demonstrated that the Marsh test was not at fault for the misleading results, but, rather, that those who performed it did it incorrectly. Orfila thus proved the presence of arsenic in LaFarge's body using the test. As a result of this, Marie was found guilty and sentenced to life imprisonment.

Effects

The case proved to be controversial, for it divided the country into factions who were convinced or otherwise of Mme. LaFarge's guilt; nevertheless, the impact of the Marsh test was great. The French press covered the trial and gave the test the publicity it needed to give the field of forensic toxicology the legitimacy it deserved, although in some ways it trivialized it: actual Marsh test assays were conducted in salons, public lectures and even in some plays that recreated the LaFarge case.

The existence of the Marsh test also served a deterrent effect: deliberate arsenic poisonings became rarer because the fear of discovery became more prevalent.

Up-and-down Procedure

The "up-and-down" procedure involves dosing animals one at a time: First one animal at one dose, then another animal one or two days later at a higher dose (if the first animal survives) or a lower dose (if the first animal dies). This process continues until the approximate LD50 has been determined. One disadvantage to this test is the length of the study. Each animal should be observed for at least seven days after dosing so that delayed deaths can be recorded. However, this method usually requires only six or eight test animals as compared with the 40 to 50 test animals that may be used in the "classical" LD50 test.

References

- Meinert CL, Tonascia S (1986). Clinical trials: design, conduct, and analysis. Oxford University Press, USA. p. 3. ISBN 978-0-19-503568-1

- O'Rourke MF (February 1992). "Frederick Akbar Mahomed". Hypertension. 19 (2): 212–7. doi:10.1161/01.HYP.19.2.212. PMC 2308176. PMID 1737655

- What-animal-testing: aboutanimaltesting.co.uk, Retrieved 12 June 2018

- Brennan Z (2013-06-05). "CROs Slowly Shifting to Adaptive Clinical Trial Designs". Outsourcing-pharma.com. Retrieved 2014-01-05

- Webb JE, Crossley MJ, Turner P, Thordarson P (June 2007). "Pyromellitamide aggregates and their response to anion stimuli". Journal of the American Chemical Society. 129 (22): 7155–62. doi:10.1021/ja0713781. PMID 17497782

- How-to-avoid-unnecessary-testing-on-animals/in-vitro-methods: echa.europa.eu, Retrieved 20 March 2018

- Simon, Harvey B. (2002). The Harvard Medical School guide to men's health. New York: Free Press. p. 31. ISBN 978-0-684-87181-3

- Helene S (2010). "EU Compassionate Use Programmes (CUPs): Regulatory Framework and Points to Consider before CUP Implementation". Pharm Med. 24 (4): 223–229. doi:10.1007/BF03256820. Archived from the original on 2012-07-07

- What-is-the-draize-test: neavs.org, Retrieved 20 May 2018

- "Adaptive Clinical Trials for Overcoming Research Challenges". News-medical.net. 2013-09-17. Retrieved 2014-01-04

- Moynihan, R. (2003). "Who pays for the pizza? Redefining the relationships between doctors and drug companies. 2: Disentanglement". British Medical Journal. 326 (7400): 1193–1196. doi:10.1136/bmj.326.7400.1193. PMC 1126054. PMID 12775622

Chapter 5

Toxicity, Indications and Side Effects

The degree to which a chemical substance can damage a living organism is known as its toxicity. It is influenced by a variety of factors, such as the pathway of administration, the duration of exposure, the number of exposures, etc. This chapter discusses the diverse aspects of toxicity, its symptoms and manifestations, through an analysis of acute, subchronic, chronic and developmental toxicity.

Toxicity

Toxicity is the degree to which a substance (a toxin or poison) can harm humans or animals.

Acute toxicity involves harmful effects in an organism through a single or short-term exposure. Subchronic toxicity is the ability of a toxic substance to cause effects for more than one year but less than the lifetime of the exposed organism. Chronic toxicity is the ability of a substance or mixture of substances to cause harmful effects over an extended period, usually upon repeated or continuous exposure, sometimes lasting for the entire life of the exposed organism.

Toxic substances can be:

- Chemical: Found naturally in the environment (for example, lead), manufactured by humans (for example, dioxin) or made by organisms (for example, chemicals made by fungi that act as toxins to insects, snake venom, or poison ivy).

- Biological: An infection by bacteria, viruses or fungi can be toxic to cells or an organism.

- Physical: The toxic effect of radiation heat and cold.

Measurement of Toxicity

The toxicity of a substance can be measured by using a bioassay – a type of test that uses a living organism. In the old days, when miners went into coal mines, they would take a canary in a cage with them. If there was toxic carbon monoxide present, the canary would die, and the miners would know they had to get out of the mine. This is a very simple type of bioassay – nowadays, scientists often use cells grown in a laboratory to test how toxic a substance is.

The toxicity measured is expressed as an LD50. This is the median level of amount of toxin that kills 50% of the tested population. For a toxin such as snake venom, the LD50 would show the number of milligrams of venom, and for a biological toxin, the LD50 might be the number of bacteria you need to ingest to kill half the tested population.

Obviously, scientists can't test toxicity on humans so they test on smaller animals or on cells and an estimate is made by comparing the relative size and type of animal.

In the example of the miners using canaries as a bioassay, the bird is small, maybe 100 times smaller than a human, so a person might need 100 times the dose for it to be lethal. But because the canary is a bird and humans are mammals, this factor could be different. A bird may be 10 times more susceptible to this particular toxin than humans, so the person might, in fact, need 1000 times the dose than that which kills the canary.

Acute Toxicity

Acute toxicity refers to those adverse effects occurring following oral or dermal administration of a single dose of a substance, or multiple doses given within 24 hours, or an inhalation exposure of 4 hours.

Classification Criteria for Substances

Chemicals can be allocated to one of five toxicity categories based on acute toxicity by the oral, dermal or inhalation route according to the numeric criteria expressed as (approximate) LD50 (oral, dermal) or LC50 (inhalation) values are shown in the table below. Explanatory notes are shown in italics following the table.

Table: Acute toxicity hazard categories and (approximate) LD50/LC50 values defining the respective categories.

Exposure Route	Category 1	Category 2	Category 3	Category 4	Category 5
Oral (mg/kg)	5	50	300	2000	
Dermal (mg/kg)	50	200	1000	2000	5000 See detailed criteria in note e
Gases (ppm)[a]	100	500	2500	5000	
Vapors (mg/l)[a b c]	0.5	2.0	10	20	
Dusts and Mists (mg/l) [a d]	0.05	0.5	1.0 5		

a. Inhalation cut-off values in the table are based on 4 hour testing exposures. Conversion of existing inhalation toxicity data which has been generated according to 1 hour exposures should be by dividing by a factor of 2 for gases and vapors and 4 for dusts and mists.

b. It is recognized that saturated vapor concentration may be used as an additional element by some regulatory systems to provide for specific health and safety protection. (e.g. UN Recommendations for the Transport of Dangerous Goods).

c. For some chemicals the test atmosphere will not just be a vapor but will consist of a mixture of liquid and vapor phases. For other chemicals the test atmosphere may consist of a vapor which is near the gaseous phase. In these latter cases, classification should be based on ppm as follows: Category 1 (100 ppm), Category 2 (500 ppm), Category 3 (2500 ppm), Category 4 (5000 ppm). Work in the OECD Test Guidelines Programme should be undertaken to better define the terms "dusts", "mists" and "vapors" in relation to inhalation toxicity testing.

d. The values for dusts and mists should be reviewed to adapt to any future changes to OECD Test Guidelines with respect to technical limitation in generating, maintaining and measuring dust and mist concentrations in respirable form.

e. Criteria for Category 5 are intended to enable the identification of substances which are of relatively low acute toxicity hazard but which, under certain circumstances may present a danger to vulnerable populations. These substances are anticipated to have an oral or dermal LD_{50} in the range of 2000-5000 mg/kg or equivalent doses for other routes. The specific criteria for Category 5 are:

- The substance is classified in this Category if reliable evidence is already available that indicates the LD_{50} or (LC_{50}) to be in the range of Category 5 values or other animal studies or toxic effects in humans indicate a concern for human health or an acute nature.

- The substance is classified in this Category, through extrapolation, estimation or measurement of data, if assignment to a more hazardous category is not warranted, and:

  o Reliable information is available indicating significant toxic effects in humans;

  o Any mortality is observed when tested up to Category 4 values by the oral, inhalation, or dermal routes;

  o Where expert judgment confirms significant clinical signs of toxicity, when tested up to Category 4 values, except for diarrhea, piloerection or an ungroomed appearance,

  o Where expert judgment confirms reliable information indicating the potential for significant acute effects from other animal studies.

Recognizing the need to protect animal welfare, testing in animals in Category 5 ranges is discouraged and should only be considered when there is a strong likelihood that results of such a test would have a direct relevance for protecting human health.

Considerations

The harmonized classification system for acute toxicity has been developed in such a way as to accommodate the needs of existing systems. A basic principle set by the IOMC CG/HCCS is that "harmonization means establishing a common and coherent basis for chemical hazard classification and communication from which the appropriate elements relevant to means of transport, consumer, worker and environment protection can be selected." To that end, five categories have been included in the acute toxicity scheme.

The preferred test species for evaluation of acute toxicity by the oral and inhalation routes is the rat, while the rat or rabbit are preferred for evaluation of acute dermal toxicity. As noted by the CG/HCCS, "Test data already generated for the classification of chemicals under existing systems should be accepted when reclassifying these chemicals under the harmonized system." When experimental data for acute toxicity are available in several animal species, scientific judgment should be used in selecting the most appropriate LD50 value from among valid, well-performed tests.

Category 1, the highest toxicity category, has cut off values of 5 mg/kg by the oral route, 50 mg/kg by the dermal route, 100 ppm for gases or gaseous vapors, 0.5 mg/l for vapors, and 0.05 mg/l for dusts and mists. These toxicity values are currently used primarily by the transport sector for classification for packing groups.

Category 5 is for chemicals which are of relatively low acute toxicity but which, under certain circumstances, may pose a hazard to especially vulnerable populations. Criteria for identifying substances in Category 5 are provided in addition to the table. These substances are anticipated to have an oral or dermal LD50 value in the range 2000 - 5000 mg/kg or equivalent doses for other routes of exposure. In light of animal welfare considerations, testing in animals in Category 5 ranges is discouraged and should only be considered when there is a strong likelihood that results of such testing would have a direct relevance for protecting human health.

Special Considerations for Inhalation Toxicity

Values for inhalation toxicity are based on 4 hour tests in laboratory animals. When experimental values are taken from tests using a 1 hour exposure, they can be converted to a 4 hour equivalent by dividing the 1 hour value by a factor of 2 for gases and vapors and 4 for dusts and mists.

Units for inhalation toxicity are a function of the form of the inhaled material. Values for dusts and mists are expressed in mg/l. Values for gases are expressed in ppm. Acknowledging the difficulties in testing vapors, some of which consist of mixtures of liquid and vapors phases, the table provides values in units of mg/l. However, for those vapors which are near the gaseous phase, classification should be based on ppm. As inhalation test methods are updated, the OECD and other test guideline programs will need to define vapors in relation to mists for greater clarity.

Vapor inhalation values are intended for use in classification of acute hazard for all sectors. It is also recognized that the saturated vapor concentration of a chemical is used by the transport sector as an additional element in classifying chemicals for packing groups.

Of particular importance is the use of well articulated values in the high toxicity categories for dusts and mists. Inhaled particles between 1 and 4 microns mean mass aerodynamic diameter (MMAD) will deposit in all regions of the rat respiratory tract. This particle size range corresponds to a maximum dose of about 2 mg/l. In order to achieve applicability of animal experiments to human exposure, dusts and mists would ideally be tested in this range in rats. The cut off values in the table for dusts and mists allow clear distinctions to be made for materials with a wide range of toxicities measured under varying test conditions. The values for dusts and mists should be reviewed in the future to adapt to any future changes in OECD or other test guidelines with respect to technical limitations in generating, maintaining, and measuring dust and mist concentrations in respirable form.

Classification Criteria For Mixtures

Considerations

The criteria for substances classify acute toxicity by use of lethal dose data (tested or derived). For mixtures, it is necessary to obtain or derive information that allows the criteria to be applied to the mixture for the purpose of classification. The approach to classification for acute toxicity is tiered, and is dependent upon the amount of information available for the mixture itself and for its ingredients. The flow chart of figure below outlines the process to be followed:

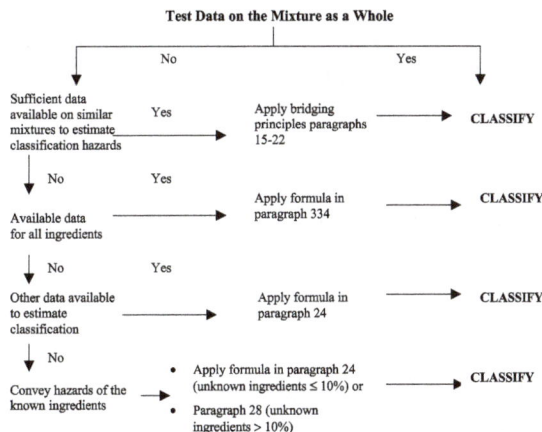

Figure: Tiered approach to classification of mixtures for acute toxicity

Classification of mixtures for acute toxicity can be carried out for each route of exposure, but is only needed for one route of exposure as long as this route is followed (estimated or tested) for all ingredients. If the acute toxicity is determined for more than one route of exposure, the more severe hazard category will be used for classification. All available information should be considered and all relevant routes of exposure should be identified for hazard communication.

In order to make use of all available data for purposes of classifying the hazards of the mixtures, certain assumptions have been made and are applied where appropriate in the tiered approach:

a) The "relevant ingredients" of a mixture are those which are present in concentrations of 1% (w/w for solids, liquids, dusts, mists and vapors and v/v for gases) or greater, unless there is a presumption that an ingredient present at a concentration of less than 1% can still be relevant for classifying the mixture for acute toxicity.

b) The acute toxicity estimate (ATE) for an ingredient in a mixture is derived using:

   • The LD50/LC50 where available,

   • The appropriate conversion value from table that relates to the results of a range test for an ingredient,

   • The appropriate conversion value from table that relates to a classification for the ingredient.

c) Where a classified mixture is used as an ingredient of another mixture, the actual or derived acute toxicity estimate (ATE) for that mixture may be used when calculating the classification of the new mixture using the formulas in paragraph 24 - 28.

Classification of Mixtures Where Acute Toxicity Test Data are Available for the Complete Mixture.

Where the mixture itself has been tested to determine its acute toxicity, it will be classified according to the criteria that have been agreed for substances. In situations where such test data for the mixture are not available, the procedures presented below should be followed.

Classification of Mixtures Where Acute Toxicity Test Data are not Available for the Complete Mixture.

Bridging Principles

Where the mixture itself has not been tested to determine its acute toxicity, but there are sufficient data on the individual ingredients and similar tested mixtures to adequately characterise the hazards of the mixture, these data will be used in accordance with the following agreed bridging rules. This ensures that the classification process uses the available data to the greatest extent possible in characterising the hazards of the mixture without the necessity for additional testing in animals.

Dilution

If a mixture is diluted with a substance that has an equivalent or lower toxicity classification than the least toxic original ingredient, and which is not expected to affect the toxicity of other ingredients, then the new mixture may be classified as equivalent to the original mixture. Alternatively, the formula explained in paragraph 24 could be applied.

If a mixture is diluted with water or other totally non-toxic material, the toxicity of the mixture can be calculated from test data on the undiluted mixture. For example, if a mixture with an LD50 of 1000 mg/kg were diluted with an equal volume of water, the LD50 of the diluted mixture would be 2000 mg/kg.

Batching

The toxicity of one production batch of a concentration mixture can be assumed to be substantially equivalent to that of another production batch of the same commercial product, and produced by or under the control of the same manufacturer, unless there is reason to believe there is significant variation such that the toxicity of the batch has changed. If the latter occurs, new classification is necessary.

Concentration of Highly Toxic Mixtures

If a mixture is classified in Category 1, and the concentration of the ingredients of the mixture that are in Category 1 is increased, the new mixture should be classified in Category 1 without additional testing.

Interpolation within one Toxicity Category

For three mixtures with identical ingredients, where A and B are in the same toxicity category and mixture C has toxicologically active ingredients with concentrations intermediate to those in mixtures A and B, then mixture C is assumed to be in the same toxicity category as A and B.

Substantially Similar Mixtures

Given the following:

a) Two mixtures

    i. A + B

    ii. C + B

b) The concentration of ingredient B is essentially the same in both mixtures.

c) The concentration of ingredient A in mixture (i) equals that of ingredient C in mixture (ii).

d) Data on toxicity for A and C are available and substantially equivalent, i.e. they are in the same hazard category and are not expected to affect the toxicity of B.

If mixture (i) is already classified by testing, mixture (ii) can be assigned the same hazard category.

Aerosols

An aerosol form of a mixture may be classified in the same hazard category as the tested, non aerosolized form of the mixture for oral and dermal toxicity provided the added propellant does not affect the toxicity of the mixture on spraying. Classification of aerosolized mixtures for inhalation toxicity should be considered separately.

Classification of Mixtures Based on Ingredients of the Mixture (Additivity Formula).

Data Available for all Ingredients

In order to ensure that classification of the mixture is accurate, and that the calculation need only be performed once for all systems, sectors, and categories, the acute toxicity estimate (ATE) of ingredients should be considered as follows:

*   Include ingredients with a known acute toxicity, which fall into any of the GHS acute toxicity categories.

*   Ignore ingredients that are presumed not acutely toxic (e.g., water, sugar).

*   Ignore ingredients if the oral limit test does not show acute toxicity at 2,000 mg/kg/body weight.

Ingredients that fall within the scope of this paragraph are considered to be ingredients with a known acute toxicity estimate (ATE).

The ATE of the mixture is determined by calculation from the ATE values for all relevant ingredients according to the following formula below for Oral, Dermal or Inhalation Toxicity:

$$\frac{100}{ATE_{mix}} = \sum_{\eta} \frac{C_i}{ATE_i}$$

Where,

C_i = concentration of ingredient i

n ingredients and i is running from 1 to n

ATE_i = Acute Toxicity Estimate of ingredient i

Where an ATE is not available for an individual ingredient of the mixture, but available information such as listed below can provide a derived conversion value, the formula in paragraph 24 may be applied.

This may include evaluation of:

a. Extrapolation between oral, dermal and inhalation acute toxicity estimates. Such an evaluation could require appropriate pharmacodynamic and pharmacokinetic data;

b. Evidence from human exposure that indicates toxic effects but does not provide lethal dose data;

c. Evidence from any other toxicity tests/assays available on the substance that indicates toxic acute effects but does not necessarily provide lethal dose data;

d. Data from closely analogous substances using structure/activity relationships.

This approach generally requires substantial supplemental technical information, and a highly trained and experienced expert, to reliably estimate acute toxicity. In the event that an ingredient without any useable information at all is used in a mixture at a concentration of 1% or greater, it is concluded that the mixture cannot be attributed a definitive acute toxicity estimate. In this situation the mixture should be classified based on the known ingredients only, with the additional statement that x percent of the mixture consists of ingredients of unknown toxicity.

If the total concentration of the ingredients with unknown acute toxicity is £ 10% then the formula presented in paragraph 24 should be used. If the total concentration of the ingredients with unknown toxicity is >10%, the formula presented in paragraph 24 should be corrected to adjust for the total percentage of the unknown ingredients as follows:

$$\frac{100 - (\Sigma C_{unknown} \text{ if } > 10\%)}{ATE_{mix}} \sum_{\eta} \frac{C_i}{ATE_i}$$

Table: Conversion from the experimentally obtained acute toxicity range estimates or a classification to point estimates for the respective routes of exposure.

	Classification or experimentally obtained acute toxicity range estimate	Conversion value
Oral (mg/kg)	0 < Category 1 ≤ 5	0.5
	5 < Category 2 ≤ 50	5
	50 < Category 3 ≤ 300	100
	300 < Category 4 ≤ 2000	500
	2000 < Category 5 ≤ 5000	2500

Dermal (mg/kg)	0 < Category 1 ≤ 50	5
	50 < Category 2 ≤ 200	50
	200 < Category 3 ≤ 1000	300
	1000 < Category 4 ≤ 2000	1100
	2000 < Category 5 ≤ 5000	2500
Gases (ppm)	0 < Category 1 ≤ 100	10
	100 < Category 2 ≤ 500	100
	500 < Category 3 ≤ 2500	700
	2500 < Category 4 ≤ 5000	3000
Vapors (mg/l)	0 < Category 1 ≤ 0.5	0.05
	0.5 < Category 2 ≤ 2.0	0.5
	2.0 < Category 3 ≤ 10.0	3
	10.0 < Category 4 ≤ 20.0	11
	Category 5	
Dust/mist (mg/l)	0 < Category 1 ≤ 0.05	0.005
	0.05 < Category 2 ≤ 0.5	0.05
	0.5 < Category 3 ≤ 1.0	0.5
	1.0 < Category 4 ≤ 5.0	1.5
	Category 5	

- Category 5 is for mixtures which are of relatively low acute toxicity but which under certain circumstances may pose a hazard to vulnerable populations. These mixtures are anticipated to have an oral or dermal LD50 value in the range of 2000-5000mg/kg or equivalent dose for other routes of exposure. In light of animal welfare considerations, testing in animals in Category 5 ranges is discouraged and should only be considered when there is a strong likelihood that results of such testing would have a direct relevance for protecting human health.

- These values are designed to be used in the calculation of the ATE for a mixture based on its components and do not represent test results. The values are conservatively set at the lower end of the range of Categories 1 and 2, and at a point approximately 1/10th from the lower end of the range for Categories 3 – 5.

Hazard Communication

Allocation of Label Elements

General and specific considerations concerning labeling requirements are provided. Annex 3 contains examples of precautionary statements and pictograms which can be used where allowed by the competent authority. Additional reference sources providing advice on the use of precautionary information is also included.

Table: Acute Toxicity Label Elements

	Category 1	Category 2	Category 3	Category 4	Category 5
Symbol	Skull and Crossbones	Skull and Crossbones	Skull and Crossbones	Exclamation Mark	No symbol is used
Signal Word	Danger	Danger	Danger	Warning	Warning
Hazard Statement: --Oral	Fatal if swallowed	Fatal if swallowed	Toxic if swallowed	Harmful if swallowed	May be harmful if swallowed
--Dermal	Fatal in contact with skin	Fatal in contact with skin	Toxic in contact with skin	Harmful in contact with skin	May be harmful in contact with skin
--Inhalation	Fatal if inhaled	Fatal if inhaled	Toxic if inhaled	Harmful if inhaled	May be harmful if inhaled

Additional Information for Category 5 Classification

As provided in Note to table, classification into Category 5 involves consideration of additional criteria, beyond the values contained in the classification table. When considering these criteria, the following guidance should be kept in mind.

"Reliable information" is considered to be information from tests that are conducted according to internationally recognized scientific principles.

Effects Considered to Support Classification

"Significant signs of toxicity" include:

- Significant functional changes in the central or peripheral nervous systems or other organ systems, including signs of central nervous system depression and effects on special senses (e.g., sight, hearing and sense of smell).

- Any consistent and significant adverse change in clinical biochemistry, haematology, or urinalysis parameters.

- Significant organ damage that may be noted at necropsy and subsequently seen or confirmed at microscopic examination.

- Multifocal or diffuse necrosis, fibrosis or granuloma formation in vital organs with regenerative capacity.

- Morphological changes that are potentially reversible but provide clear evidence of marked organ dysfunction.

- Evidence of appreciable cell death (including cell degeneration and reduced cell number) in vital organs incapable of regeneration.

Effects Considered not to Support Classification

It is recognized that effects may be seen that would not justify classification. Examples of such effects in humans and animals are include: clinical observations or small changes in bodyweight gain, food consumption or water intake that may have some toxicological importance but that do not, by themselves, indicate "significant" toxicity.

Decision Tree for Classification of Acute Toxicity.

Calculation

The ACR is the inverse of the application factor (AF). This makes it easier for regulators to visualize data as whole numbers rather than decimals. The AF is calculated by dividing the Maximum Acceptable Toxicant Concentration (MATC) by the Lethal Concentration that kills 50% of test organisms in an acute toxicity test (LC50).

$$MATC = \sqrt{(NOEC)(LOEC)}$$

The Maximum Allowable Toxicity Concentration (MATC is determined by taking the square root of the No Effects Concentration (NOEC) multiplied by the Low effect concentration (LOEC),

$$AF = MATC / LC50$$

The Application Factor (AF) is determined by dividing the MATC by the LC50,

$$ACR = 1 / AF \quad \text{or} \quad LC50 / MATC$$

The ACR is then the inverse of the AF.

Regulatory use

There are thousands of new and different chemicals that are designed and synthesized by private chemical manufacturers every year. The public demands that all of these chemicals go through testing and be approved for use by the EPA under the TSCA. Part of that testing requirement is determining the toxicity of chemicals to organisms in the environment.

Law

Section 5 of the TSCA states that the EPA must respond to pre-manufacturing notices (PMN) 90 –180 days after submission by the manufacturer. The EPA is responsible for identifying the substance, its proposed use, amount made, byproducts, exposure levels, and all existing environmental and health data necessary to prevent significant harm to the environment. Additionally there are no PMN test requirements so there is often a minimal amount of data presented. This may be discussed as a fault of the TSCA. New chemical PMNs are submitted early in the chemical's development so they rarely contain information about chronic toxicity - yet the EPA must respond within the 90-180 day time period after submission of the PMN. This essentially puts a huge burden on the EPA because chemical effects to the environment are extremely hard to predict simply based on single species toxicity tests (SST). The limited time period that the TSCA gives the EPA for making this decision requires the EPA to make decisions with a high amount of uncertainty. This ultimately makes the goal of protecting the environment from significant adverse effects difficult.

The results of acute and chronic toxicity testing form the basis of knowledge that regulators draw from in performing work related to ecological risk assessment and designing policy that defines how much of a chemical of interest should be allowed in certain environments. While this sounds simple enough to the layperson, it is extremely difficult in practice due to a large number of modifying factors inextricably tied to toxicity tests and statistical analysis. Different toxic effects can be observed from the same chemical through different types of environmental exposures and parameters, and thus toxicity results from acute and chronic tests must be jointly considered in decision making. Additionally, chronic toxicity tests tend to require significantly more attention and resources than acute tests which makes them much less feasible for basing decisions off of in a timely manner. The need for development of more advanced statistical methods, and uniformity in using these methods by regulators has been made apparent in literature.

Scientific methods for determining acute and chronic toxicity to organisms are inherently imperfect and non-uniform throughout the field of research, and the most useful tool for decision making by officials is more often than not best personal judgment.

A popular new method for ecological risk assessment is the acute to chronic estimation (ACE). This method uses computer software to estimate chronic toxicity, which provides similar information with much less effort and expense to the researcher.

Limitations

The ACR is derived from data generated by SSTs, as so falls victim to the same errors and limitations. These limitations are described in detail in literature

Using point estimates such as NOECs/LOECs reduces a data set containing many values down to an isometric, removing the rich visual information that allows the researcher to assess the reliability and variability in the data. Information such as the slope of the dose-response curve, from which NOECs are LOECs are derived, is lost. However, without NOECs and LOECs regulatory decisions are much harder to make. While ACR has drawbacks due to the uncertainty of the point estimates it uses to define it, it is still widely valued as a regulatory tool in making environmental assessments and policy decisions.

ACRs are based off tests with a number of different methodologies, which means that there can be significant variance among ACRs.

Chronic Toxicity

Chronic toxicity is a property of a substance that has toxic effects on a living organism, when that organism is exposed to the substance continuously or repeatedly. Compared with acute toxicity two distinct situations need to be considered:

- Prolonged exposure to a substance For example if a person drinks too much alcohol on a regular basis then their health may suffer as a result. The alcohol does not have a long biological halflife but it is supplied on a regular basis to the body of the person.

- Prolonged internal exposure because a substance remains in the body for a long time For example if a person were to ingest radium much of it would be absorbed into the bones where it would exert a harmful effect on a person's health. The radium might cause a disturbance in the blood cell-forming part of the bone.

The objective of a chronic toxicity study is to determine the effects of a substance in a mammalian species following prolonged and repeated exposure. Under the conditions of the chronic toxicity test, effects which require a long latency period or which are cumulative should become manifest. The application of this guideline should generate data on which to identify the majority of chronic effects and shall serve to define long term dose-response relationships. The design and conduct of chronic toxicity tests should allow for the detection of general toxic effects, including neurological, physiological, biochemical, and hematological effects and exposure-related morphological (pathology) effects.

Test Procedures

1) *Animal selection*

   i   *Species and strain:* Testing should be performed with two mammalian species, one a rodent and another a non-rodent. The rat is the preferred rodent species and the dog is the preferred non-rodent species. Commonly used laboratory strains should be employed. If other mammalian species are used, the tester should provide justification/reasoning for their selection.

   ii) *Age*

      A) Dosing of rats should begin as soon as possible after weaning, ideally before the rats are 6, but in no case more than 8 weeks old.

    B) Dosing of dogs should begin between 4 and 6 months of age and in no case later than 9 months of age.

    C) At commencement of the study the weight variation of animals used should not exceed ±20 percent of the mean weight for each sex.

iii) Sex

    A) Equal numbers of animals of each sex should be used at each dose level.

    B) The females should be nulliparous and non-pregnant.

iv) Numbers

    A) For rodents, at least 40 animals (20 females and 20 males) and for non-rodents (dogs) at least 8 animals (4 females and 4 males) should be used at each dose level.

    B) If interim sacrifices are planned, the number should be increased by the number of animals scheduled to be sacrificed during the course of the study.

    C) The number of animals at the termination of the study must be adequate for a meaningful and valid statistical evaluation of chronic effects.

2) *Control groups*

    i) A concurrent control group is suggested. This group should be an untreated or sham treated control group or, if a vehicle is used in administering the test substance, a vehicle control group. If the toxic properties of the vehicle are not known or cannot be made available, both untreated and vehicle control groups are strongly suggested.

    ii) In special circumstances such as in inhalation studies involving aerosols or the use of an emulsifier of uncharacterized biological activity in oral studies, a concurrent negative control group should be utilized. The negative control group should be treated in the same manner as all other test animals except that this control group should not be exposed to either the test substance or any vehicle.

3) *Dose levels and dose selections*

    i) In chronic toxicity tests, it is necessary to have a dose-response relationship as well as a no-observed-toxic-effect level. Therefore, at least three dose levels should be used in addition to the concurrent control group. Dose levels should be spaced to produce a gradation of effects.

    ii) The high dose level in rodents should elicit some signs of toxicity without causing excessive lethality; for non-rodents, there should be signs of toxicity but there should be no fatalities.

iii) The lowest dose level should not produce any evidence of toxicity. Where there is a usable estimation of human exposure the lowest dose level should exceed this even though this dose level may result in some signs of toxicity.

iv) Ideally, the intermediate dose level(s) should produce minimal observable toxic effects. If more than one intermediate dose is used, the dose level should be spaced to produce a gradation of toxic effects.

v) For rodents, the incidence of fatalities in low and intermediate dose groups and in the controls should be low to permit a meaningful evaluation of the results. For non-rodents, there should be no fatalities.

4) *Exposure conditions*

The animals are dosed with the test substance ideally on a 7-day per week basis over a period of at least 12 months. However, based primarily on practical considerations, dosing on a 5-day per week basis is considered to be acceptable.

5) *Observation period*

Duration of observation should be for at least 12 months, and may be concurrent with or subsequent to dosing. If there is a post-exposure observation period, an interim sacrifice should be performed on no fewer than half of the animals of each sex at each dose level immediately upon termination of exposure.

6) *Administration of the test substance*

The three main routes of administration are oral, dermal, and inhalation. The choice of the route of administration depends upon the physical and chemical characteristics of the test substance and the form typifying exposure in humans.

i) *Oral studies*

A) The animals should receive the test substance in their diet, dissolved in drinking water, or given by gavage or capsule for a period of at least 12 months.

B) If the test substance is administered in the drinking water, or mixed in the diet, exposure is continuous.

C) For a diet mixture, the highest concentration should not exceed 5 percent.

ii) *Dermal studies*

A) The animals are treated by topical application with the test substance, ideally for at least 6 hours per day.

B) Fur should be clipped from the dorsal area of the trunk of the test animals. Care must be taken to avoid abrading the skin which could alter its permeability.

C) The test substance should be applied uniformly over a shaved area which is approximately 10 percent of the total body surface area. With highly toxic substances, the surface area covered may be less, but as much of the area should be covered with as thin and uniform a film as possible.

D) During the exposure period, the test substance may be held if necessary, in contact with the skin with a porous gauze dressing and non-irritating tape. The test site should be further covered in a suitable manner to retain the gauze dressing and test substance and ensure that the animals cannot ingest the test substance.

iii) *Inhalation studies*

A) The animals should be tested with inhalation equipment designed to sustain a dynamic air flow of 12 to 15 air changes per hour, ensure an adequate oxygen content of 19 percent and an evenly distributed exposure atmosphere. Where a chamber is used, its design should minimize crowding of the test animals and maximize their exposure to the test substance. This is best accomplished by individual caging. As a general rule to ensure stability of a chamber atmosphere, the total "volume" of the test animals should not exceed 5 percent of the volume of the test chamber. Alternatively, oro-nasal, head-only or whole body individual chamber exposure may be used.

B) The temperature at which the test is performed should be maintained at 22° C (±2°). Ideally, the relative humidity should be maintained between 40 to 60 percent, but in certain instances (e.g., tests of aerosols, use of water vehicle) this may not be practicable.

C) Feed and water should be withheld during each daily 6 hour exposure period.

D) A dynamic inhalation system with a suitable analytical concentration control system should be used. The rate of air flow should be adjusted to ensure that conditions throughout the equipment are essentially the same. Maintenance of slight negative pressure inside the chamber will prevent leakage of the test substance into the surrounding areas.

7) *Observation of animals*

i) Each animal should be handled and its physical condition appraised at least once each day.

ii) Additional observations should be made daily with appropriate actions taken to minimize loss of animals to the study (e.g., necropsy or refrigeration of those animals found dead and isolation or sacrific of weak or moribund animals).

iii) Clinical signs of toxicity including suspected tumors and mortality should be recorded as they are observed, including the time of onset, the degree and duration.

iv) Cage-side observations should include, but not be limited to, changes in skin and fur, eyes and mucous membranes, respiratory, circulatory, autonomic and central nervous systems, somatomotor activity and behavior pattern.

v) Body weights should be recorded individually for all animals once a week during the first 13 weeks of the test period and at least once every 4 weeks thereafter unless signs of clinical toxicity suggest more frequent weighings to facilitate monitoring of health status.

vi) When the test substance is administered in the feed or drinking water, measurements of feed or water consumption, respectively, should be determined weekly during the first 13 weeks of the study and then at approximately monthly intervals unless health status or body weight changes dictate otherwise.

vii) At the end of the study period all survivors should be sacrificed. Moribund animals should be removed and sacrificed when noticed.

8) *Physical measurements*

For inhalation studies, measurements or monitoring should be made of the following:

i) The rate of air flow should be monitored continuously, but should be recorded at intervals of at least once every 30 minutes.

ii) During each exposure period the actual concentrations of the test substance should be held as constant as practicable, monitored continuously and measured at least three times during the test period: at the beginning, at an intermediate time and at the end of the period.

iii) During the development of the generating system, particle size analysis should be performed to establish the stability of aerosol concentrations. During exposure, analysis should be conducted as often as necessary to determine the consistency of particle size distribution and homogeneity of the exposure stream.

iv) Temperature and humidity should be monitored continuously, but should be recorded at intervals of at least once every 30 minutes.

9) *Clinical examinations*

The following examinations should be made on at least 10 rats of each sex per dose and on all non-rodents.

i) Certain hematology determinations (e.g., hemoglobin content, packed cell volume, total red blood cells, total white blood cells, platelets, or other measures of clotting potential) should be performed at termination and should be performed at 3 months, 6 months and at approximately 6 month intervals thereafter (for studies extending beyond 12 months) on blood samples collected from all non-rodents and from 10 rats per sex of all groups.

These collections should be from the same animals at each interval. If clinical observations suggest a deterioration in health of the animals during the study, a differential blood count of the affected animals should be performed. A differential blood count should be performed on samples from those animals in the highest dosage group and the controls. Differential blood counts should be performed for the next lower group(s) if there is a major discrepancy between the highest group and the controls. If hematological effects were noted in the subchronic test, hematological testing should be performed at 3, 6, 12, 18, and 24 months for a two year study and at 3, 6, and 12 months for a 1-year study.

ii) Certain clinical biochemistry determinations on blood should be carried out at least three times during the test period: just prior to initiation of dosing (base line data), near the middle and at the end of the test period. Blood samples should be drawn for clinical chemistry measurements from all non-rodents and at least ten rodents per sex of all groups; if possible, from the same rodents at each time interval. Test areas which are considered appropriate to all studies: electrolyte balance, carbohydrate metabolism and liver and kidney function. The selection of specific tests will be influenced by observations on the mode of action of the substance and signs of clinical toxicity. Suggested chemical determinations: calcium, phosphorus, chloride, sodium, potassium, fasting glucose (with period of fasting appropriate to the species), serum glutamic-pyruvic transaminase (now known as serum alanine aminotransferase), serum glutamic oxaloacetic transaminase (now known as serum aspartate aminotransferase), ornithine decarboxylase, gamma glutamyl transpeptidase, blood urea nitrogen, albumen, blood creatinine, creatinine phosphokinase, total cholesterol, total bilirubin and total serum protein measurements. Other determinations which may be necessary for an adequate toxicological evaluation include analyses of lipids, hormones, acid/base balance, methemoglobin and cholinesterase activity. Additional clinical biochemistry may be employed where necessary to extend the investigation of observed effects.

iii) Urine samples from rodents at the same intervals as the hematological examinations under paragraph (b)(9)(i) of this topic should be collected for analysis. The following determinations should be made from either individual animals or on a pooled sample/sex/group for rodents: appearance (volume and specific gravity), protein, glucose, ketones, bilirubin, occult blood (semi-quantitatively); and microscopy of sediment (semi-quantitatively).

iv) Ophthalmological examination, using an ophthalmoscope or equivalent suitable equipment, should be made prior to the administration of the test substance and at the termination of the study. If changes in eyes are detected all animals should be examined.

10) *Gross necropsy*

   i) A complete gross examination should be performed on all animals, including those which died during the experiment or were killed in moribund conditions.

   ii) The liver, kidneys, adrenals, brain and gonads should be weighed wet, as soon as possible after dissection to avoid drying. For these organs, at least 10 rodents per sex per group and all non-rodents should be weighed.

   iii) The following organs and tissues, or representative samples thereof, should be preserved in a suitable medium for possible future histopathological examination: All gross lesions and tumors; brain-including sections of medulla/pons, cerebellar cortex, and cerebral cortex; pituitary; thyroid/parathyroid; thymus; lungs; trachea; heart; sternum and femur with bone marrow; salivary glands; liver; spleen; kidneys; adrenals; esophagus; stomach; duodenum; jejunum; ileum; cecum; colon; rectum; urinary bladder; representative lymph nodes; pancreas; gonads; uterus; accessory genital organs (epididymis, prostate, and, if present, seminal vesicles; female mammary gland; aorta; gall bladder (if present); skin; musculature; peripheral nerve; spinal cord at three levels - cervical, midthoracic, and lumbar; and eyes. In inhalation studies, the entire respiratory tract, including nose, pharynx, larynx, and paranasal sinuses should be examined and preserved. In dermal studies, skin from sites of skin painting should be examined and preserved.

   iv) Inflation of lungs and urinary bladder with a fixative is the optimal method for preservation of these tissues. The proper inflation and fixation of the lungs in inhalation studies is considered essential for appropriate and valid histopathological examination.

   v) If other clinical examinations are carried out, the information obtained from these procedures should be available before microscopic examination, since they may provide significant guidance to the pathologist.

11) *Histopathology*

   i) The following histopathology should be performed:

      A) Full histopathology on the organs and tissues, listed above, of all non-rodents, of all rodents in the control and high dose groups and of all rodents that died or were killed during the study.

      B) All gross lesions in all animals.

      C) Target organs in all animals.

      D) Lungs, liver and kidneys of all animals. Special attention to examination of the lungs of rodents should be made for evidence of infection since this provides an assessment of the state of health of the animals.

ii) If excessive early deaths or other problems occur in the high dose group compromising the significance of the data, the next dose level should be examined for complete histopathology.

iii) In case the results of an experiment give evidence of substantial alteration of the animals' normal longevity or the induction of effects that might affect a toxic response, the next lower dose level should be examined fully, as described under paragraph (b)(11)(i) of this topic.

iv) An attempt should be made to correlate gross observations with microscopic findings.

Data and Reporting

1) *Treatment of results*

i) Data should be summarized in tabular form, showing for each test group the number of animals at the start of the test, the number of animals showing lesions, the types of lesions and the percentage of animals displaying each type of lesion.

ii) All observed results, quantitative and incidental, should be evaluated by an appropriate statistical method. Any generally accepted statistical methods may be used; the statistical methods should be selected during the design of the study.

2) *Evaluation of study results*

i) The findings of a chronic toxicity study should be evaluated in conjunction with the findings of preceding studies and considered in terms of the toxic effects, the necropsy and histopathological findings. The evaluation will include the relationship between the dose of the test substance and the presence, incidence and severity of abnormalities (including behavioral and clinical abnormalities), gross lesions, identified target organs, body weight changes, effects on mortality and any other general or specific toxic effects.

ii) In any study which demonstrates an absence of toxic effects, further investigation to establish absorption and bioavailability of the test substance should be considered.

(3) *Test report*

i) In addition to the reporting requirements as specified under 40 CFR part 792 subpart J, the following specific information should be reported:

A) *Group animal data.* Tabulation of toxic response data by species, strain, sex and exposure level for:

1) Number of animals dying.

2) Number of animals showing signs of toxicity.

3) Number of animals exposed.

B) *Individual animal data.*

1) Time of death during the study or whether animals survived to termination.

2) Time of observation of each abnormal sign and its subsequent course.

3) Body weight data.

4) Feed and water consumption data, when collected.

5) Results of ophthalmological examination, when performed.

6) Hematological tests employed and all results.

7) Clinical biochemistry tests employed and all results.

8) Necropsy findings.

9) Detailed description of all histopathological findings.

10) Statistical treatment of results, where appropriate.

ii) In addition, for inhalation studies the following should be reported:

A) *Test conditions*

1) Description of exposure apparatus including design, type, dimensions, source of air, system for generating particulates and aerosols, method of conditioning air, treatment of exhaust air and the method of housing the animals in a test chamber.

2) The equipment for measuring temperature, humidity, and particulate aerosol concentrations and size should be described.

B) *Exposure data.* These should be tabulated and presented with mean values and a measure of variability (e.g., standard deviation) and should include:

1) Airflow rates through the inhalation equipment.

2) Temperature and humidity of air.

3) Nominal concentration (total amount of test substance fed into the inhalation equipment divided by volume of air).

4) Actual concentration in test breathing zone.

5) Particle size distribution (e.g., median aerodynamic diameter of particles with standard deviation from the mean).

Carcinogenicity

The term "carcinogen" denotes a chemical substance or a mixture of chemical substances which induce cancer or increase its incidence. Substances which have induced benign and malignant tumors in well performed experimental studies on animals are considered also to be presumed or suspected human carcinogens unless there is strong evidence that the mechanism of tumor formation is not relevant for humans.

Classification of a chemical as posing a carcinogenic hazard is based on the inherent properties of the substance and does not provide information on the level of the human cancer risk which the use of the chemical may represent.

Considerations

The purpose of the harmonized system for the classification of chemicals which may cause cancer is to provide common ground which could be used internationally for the classification of carcinogenic substances.

The scheme is applicable to the classification of all chemicals. Its application to classification of mixtures is explained in paragraphs 13-18.

Classification Criteria for Substances

For the purpose of classification for carcinogenicity, chemical substances are allocated to one of two classes based on strength of evidence and additional considerations (weight of evidence). In certain instances route specific classification may be warranted.

Category 1: Known or Presumed Human Carcinogens

The placing of a chemical in Category 1 is done on the basis of epidemiological and animal data. An individual chemical may be further distinguished:

Category 1A: Known to have carcinogenic potential for humans; the placing of a chemical is largely based on human evidence.

Category 1B: Presumed to have carcinogenic potential for humans; the placing of a chemical is largely based on animal evidence.

Based on strength of evidence together with additional considerations, such evidence may be derived from human studies that establish a causal relationship between human exposure to a chemical and the development of cancer (known human carcinogen). Alternatively, evidence may be derived from animal experiments for which there is sufficient evidence to demonstrate animal carcinogenicity (presumed human carcinogen). In addition, on a case by case basis, scientific judgement may warrant a decision of presumed human carcinogenicity derived from studies showing limited evidence of carcinogenicity in humans together with limited evidence of carcinogenicity in experimental animals.

Classification: Category 1 (A and B) Carcinogen

Category 2: Suspected Human Carcinogens

The placing of a chemical in Category 2 is done on the basis of evidence obtained from human and animal studies, but which is not sufficiently convincing to place the chemical in Category 1.

Based on strength of evidence together with additional considerations, such evidence may be from either limited evidence of carcinogenicity in human studies or from limited evidence of carcinogenicity in animal studies.

Classification: Category 2 Carcinogen

Rationale

Classification as Carcinogen is made on the basis of evidence from reliable and acceptable methods, and is intended to be used for chemicals which have an intrinsic property to produce such toxic effects. The evaluations should be based on all existing data, peer-reviewed published studies and additional data accepted by regulatory agencies.

Carcinogen classification is a one-step, criterion-based process that involves two interrelated determinations: evaluations of strength of evidence and consideration of all other relevant information to place chemicals with human cancer potential into hazard classes.

Strength of evidence involves the enumeration of tumors in human and animal studies and determination of their level of statistical significance. Sufficient human evidence demonstrates causality between human exposure and the development of cancer, whereas sufficient evidence in animals shows a causal relationship between the agent and an increased incidence of tumors. Limited evidence in humans is demonstrated by a positive association between exposure and cancer, but a causal relationship cannot be stated. Limited evidence in animals is provided when data suggest a carcinogenic effect, but are less than sufficient. The terms "sufficient" and "limited" are used here as they have been defined by the International Agency for Research on Cancer (IARC) and are cited in the Background Information for this document.

Additional considerations (weight of evidence). Beyond the determination of the strength of evidence for carcinogenicity, a number of other factors should be considered that influence the overall likelihood that an agent may pose a carcinogenic hazard in humans. The full list of factors that influence this determination is very lengthy, but some of the important ones are considered here.

The factors can be viewed as either increasing or decreasing the level of concern for human carcinogenicity. The relative emphasis accorded to each factor depends upon the amount and coherence of evidence bearing on each. Generally there is a requirement for more complete information to decrease than to increase the level of concern. Additional considerations should be used in evaluating the tumor findings and the other factors in a case-by-case manner.

Some important factors which may be taken into consideration, when assessing the overall level of concern are:

- Tumor type and background incidence.

- Multisite responses.

- Progression of lesions to malignancy.

- Reduced tumor latency.

Additional factors on which the evaluation may increase or decrease the level of concern include:

- Whether responses are in single or both sexes.

- Whether responses are in a single species or several species.

- Structural similarity or not to a chemical(s) for which there is good evidence of carcinogenicity.

- Routes of exposure.

- Comparison of absorption, distribution, metabolism and excretion between test animals and humans.

- The possibility of a confounding effect of excessive toxicity at test doses.

- Mode of action and its relevance for humans, such as mutagenicity, cytotoxicity with growth stimulation, mitogenesis, immunosuppression.

Mutagenicity: It is recognized that genetic events are central in the overall process of cancer development. Therefore evidence of mutagenic activity in vivo may indicate that a chemical has a potential for carcinogenic effects.

Classification Criteria for Mixtures

Classification of mixtures will be based on the available test data of the individual constituents of the mixture using cut-off values/concentration limits for the components of the mixture. The classification may be modified on a case-by case basis based on the available test data for the mixture as a whole. In such cases, the test results for the mixture as a whole must be shown to be conclusive taking into account dose and other factors such as duration, observations and analysis (e.g., statistical analysis, test sensitivity) of carcinogenicity test systems. Adequate documentation supporting the classification should be retained and made available for review upon request.

Bridging Principles

Where the mixture itself has not been tested to determine its carcinogenic hazard, but there are sufficient data on the individual ingredients and similar tested mixtures to adequately characterize the hazards of the mixture, this data will be used in accordance with the following agreed bridging rules. This ensures that the classification process uses the available data to the greatest extent possible in characterizing the hazards of the mixture without the necessity for additional testing in animals.

Dilution

If a mixture is diluted with a diluent which is not expected to affect the carcinogenicity of other ingredients, then the new mixture may be classified as equivalent to the original mixture.

Batching

The carcinogenic potential of one production batch of a complex mixture can be assumed to be substantially equivalent to that of another production batch of the same commercial product produced by and under the control of the same manufacture unless there is reason to believe there is significant variation in composition such that the carcinogenic potential of the batch has changed. If the latter occurs, a new classification is necessary.

Substantially Similar Mixtures

Given the following:

a) Two mixtures

   i.) A + B

   ii.) C + B

b) The concentration of carcinogen ingredient B is the same in both mixtures.

c) The concentration of ingredient A in mixture i equals that of ingredient C in mixture ii.

d) Data on toxicity for A and C are available and substantially equivalent, i.e. they are not expected to affect the carcinogenicity of B.

If mixture (i) is already classified by testing, mixture (ii) can be assigned the same category.

Classification of Mixtures When Data are Available for All Components or Only for Some Components of the Mixture.

The mixture will be classified as a carcinogen when at least one ingredient has been classified as a Category 1 or Category 2 carcinogen and is present at or above the appropriate cut-off value/concentration limit as mentioned in table below for Category 1 and 2 respectively.

Table: Cut-off values/concentration limits of ingredients of a mixture classified as carcinogen that would trigger classification of the mixture.

Ingredient classified as:	Cut-off/concentration limits triggering classification of a mixture as:	
	Category 1 carcinogen	Category 2 carcinogen
Category 1 carcinogen	0.1 %	
Category 2 carcinogen	-	≥ 0.1% (note1)
		≥ 1.0% (note 2)

Note 1: If a Category 2 carcinogen ingredient is present in the mixture at a concentration between 0.1% and 1%, every regulatory authority would require information on the MSDS for a product. However, a label warning would be optional. Some authorities will choose to label when the ingredient is present in the mixture between 0.1% and 1%, whereas others would normally not require a label in this case.

Note 2: If a Category 2 carcinogen ingredient is present in the mixture at a concentration of ≥ 1%, both an MSDS and a label would generally be expected.

Hazard Communication

Allocation of Label Elements

Annex 5 contains examples of precautionary statements and pictograms which can be used where allowed by the competent authority. Additional reference sources providing advice on the use of precautionary information is also included.

	Category 1A	Category 1B	Category 2
Symbol	New health hazard symbol	New health hazard symbol	New health hazard symbol
Signal Word	Danger	Danger	Warning
Hazard Statement	May cause cancer (state route of exposure if it is conclusively proven that no other routes of exposure cause the hazard)	May cause cancer (state route of exposure if it is conclusively proven that no other routes of exposure cause the hazard)	Suspected of causing cancer (state route of exposure if it is conclusively proven that no other routes of exposure cause the hazard)

Developmental Toxicity

Developmental toxicity refers to adverse effects on the developing organism that result from exposure prior to conception, during the prenatal period, or postnatally up to the time of sexual maturity. The four major manifestations of developmental toxicity are:

- Mortality

- Dysmorphogenesis (structural abnormalities)

- Alterations to growth

- Functional impairment

Mortality

Mortality resulting from developmental toxicity may occur at any time from early conception to post-weaning (e.g., embryo-fetal death is a subset of mortality resulting from developmental toxicity). Thus, a positive signal may appear as:

- Pre- or peri-implantation loss

- Early or late resorption

- Abortion

- Stillbirth

- Neonatal death

- Peri-weaning loss

Dysmorphogenesis

Dysmorphogenic effects are generally seen as malformations or variations of the skeleton or soft tissues of the offspring and are commonly referred to as structural abnormalities.

Alterations in Growth

Alterations in growth are generally seen as growth retardation, although excessive growth or early maturation may also be considered alterations to growth. Body weight is the most common measurement for assessing growth rate. Crown-rump length and anogenital distance may also be measured. Sometimes it is not clear if an effect is a direct structural alteration or an inhibition of growth. For example, reduced ossification could be either. A distinction must be made upon review of all relevant data.

Functional Impairment

Functional toxicities could include any persistent alteration of normal physiologic or biochemical function, but typically only developmental neurobehavioral effects and reproductive function are measured. Common assessments include:

- Locomotor activity
- Learning and memory
- Reflex development
- Time to sexual maturation
- Mating behavior
- Fertility

The process by which a toxicant can produce dysmorphogenesis, growth retardation, lethality, and functional alterations commonly is referred to as the "mechanism" by which developmental toxicity is produced. In general, it has been difficult to analyze mechanisms in sufficient detail and depth for risk assessment purposes. There are four reasons.

1. Normal development is extremely complex, and it is possible that there is a myriad of points at which a toxicant might interact with an important molecular component and cause developmental toxicity. Information about molecular components and processes of development has only been available in the past few years, largely through the study of developmental mutants of invertebrate model organisms, such as Drosophila and Caenorhabditis elegans. As highlighted by Wilson's principles, an understanding of mechanisms would be greatly enhanced by identifying critical key events altered by toxicants. Recent advances in research on signaling pathways and genetic regulatory circuits in development might have identified especially critical processes, ones that, if studied for their alteration by developmental toxicants, might provide exciting new clues for mechanistic investigations. For now, such insights are available in only a few cases, such as toxicant interactions with components of the nuclear hormone-receptor family of signal receptors and gene regulators.

2. Environmental toxicants include a wide range of chemical, physical, and biological agents that initiate a wide variety of mechanisms. Some agents are specific for one or a few targets in the development or physiology of the conceptus, and others have a broad effect on many targets at different times and places in the conceptus and mother. Thus, the developmental toxicologist who focuses on these agents is probably faced with a wide variety of mechanisms.

3. Some toxicants might affect only a fraction of individuals in the population, probably because of genetic differences or differences in health history (diseases, nutrition, or other exposures). The differences add considerable complexity.

4. A mechanistic understanding of developmental toxicity involves understanding at several levels of biological organization. Once a toxicant interacts with a molecular component of the cell, it presumably affects its immediate function, so the function and alteration must be known. Then, the consequence for the altered function for the completion of a developmental process must be known. For example, in order to link specific branchiofacial defects with the action of a suspected toxicant, it is necessary to characterize the migratory events, proliferation control processes, and patterns of differentiation-promoting signal systems that affect neural crest cells from the time of their emigration from the neural tube. Other kinds of toxicants might alter specialized functions of organs of the fetus (e.g., the heart) and thus manifest impacts at the organ level. Yet other toxicants might cause cell death in the conceptus at a variety of times and locations and have multiple impacts.

References

• Office, U.S. Government Accountability (2 December 2009). "Chemical Regulation: Observations on Improving the Toxic Substances Control Act". Gao.gov (GAO-10–292T). Retrieved 7 January 2018

• Toxicity-1540, sciencelearn.org.nz, Retrieved 11 June 2018

• Maltby, L., Clayton, S. A., Yu, H., McLoughlin, N., Wood, R. M. and Yin, D. (2000). "Using single-species toxicity tests, community-level responses, and toxicity identification evaluations to investigate effluent impacts". Environmental Toxicology and Chemistry. 19: 151–157. doi:10.1002/etc.5620190118

• Chronic-toxicity: definitions.net, Retrieved 28 March 2018

• Fox, D. R. and Landis, W. G. (2016). "Don't be fooled—A no-observed-effect concentration is no substitute for a poor concentration–response experiment". Environ Toxicol Chem. 35: 2141–2148. doi:10.1002/etc.3459

Chapter 6

Toxic Effects of Substances

Exposure to a toxic substance can affect the health of an organism. This chapter closely examines the toxic effects of substances such as arsenic, carbon monoxide, chloride gas, fluoride, sulphuric acid, etc.

Arsenic Poisoning

Arsenic is a naturally occurring, metalloid component of the Earth's crust. Minuscule quantities of arsenic occur in all rock, air, water, and soil. A metalloid is a substance that is not a metal but shares many qualities with metals.

The concentration of arsenic may be higher in certain geographical regions. This could be a result of human activity, such as metal mining or the use of pesticides. Natural conditions can also lead to a higher concentration.

It can be found combined with other elements in different chemical compounds. Organic forms of arsenic also contain carbon, but inorganic forms do not. Arsenic cannot be dissolved in water.

Inorganic arsenic compounds are more harmful than organic ones. They are more likely to react with the cells in the body, displace certain elements from the cell, and change the cell's function.

For example, cells use phosphate for energy generation and signaling, but one form of arsenic, known as arsenate, can imitate and replace the phosphate in the cell. This impairs the ability of the cell to generate energy and communicate with other cells.

This cell-altering ability may be useful in cancer treatment, as some studies have shown it can send the disease into remission and help thin the blood. Arsenic-based chemotherapy drugs, such as arsenic trioxide, are already in use for some cancers.

Arsenic poisoning, or arsenicosis, happens when a person takes in dangerous levels of arsenic. Arsenic is a natural semi-metallic chemical that is found all over the world in groundwater.

Intake can result from swallowing, absorbing, or inhaling the chemical.

Arsenic poisoning can cause major health complications and death if it is not treated, so precautions exist to protect those who are at risk.

Arsenic is often implicated in deliberate poisoning attempts, but an individual can be exposed to arsenic through contaminated groundwater, infected soil, and rock, and arsenic-preserved wood.

However, arsenic in the environment is not immediately dangerous, and it is rare to find toxic amounts of arsenic in nature.

Fast Facts about Arsenic Poisoning

- Arsenic is a natural metalloid chemical that may be present in groundwater.

- Ingestion only poses health problems if a dangerous amount of arsenic enters the body. Then, it can lead to cancer, liver disease, coma, and death.

- Treatment involves bowel irrigation, medication, and chelation therapy.

- It is rare to find dangerous amounts of arsenic in the natural environment. Areas with dangerous arsenic levels are usually well-known and provisions exist to prevent and handle the risk of poisoning.

- Anyone who suspects there may be high arsenic levels in their local environment should contact their local authorities for more information.

Symptoms

The symptoms of arsenic poisoning can be acute, or severe and immediate, or chronic, where damage to health is experienced over a longer period. This will often depend on the method of exposure.

A person who has swallowed arsenic may show signs and symptoms within 30 minutes.

These may include:

- Drowsiness

- Headaches

- Confusion

- Severe diarrhea

If arsenic has been inhaled, or a less concentrated amount has been ingested, symptoms may take longer to develop. As the arsenic poisoning progresses, the patient may start experiencing convulsions, and their fingernail pigmentation may change.

Signs and symptoms associated with more severe cases of arsenic poisoning are:

- A metallic taste in the mouth and garlicky breath
- Excess saliva
- Problems swallowing
- Blood in the urine
- Cramping muscles
- Hair loss
- Stomach cramps
- Convulsions
- Excessive sweating
- Vomiting
- Diarrhea

Arsenic poisoning typically affects the skin, liver, lungs, and kidneys. In the final stage, symptoms include seizures and shock. This could lead to a coma or death.

Complications

Complications linked to long-term arsenic consumption include:

- Cancer
- Liver disease
- Diabetes
- Nervous system complications, such as loss of sensation in the limbs and hearing problems
- Digestive difficulties

Causes

The main cause of arsenic poisoning is the consumption of a toxic amount of arsenic.

Arsenic, consumed in large amounts, can kill a person rapidly. Consumed in smaller amounts over a long period, it can cause serious illness or a prolonged death.

The main cause of arsenic poisoning worldwide is the drinking of groundwater that contains high levels of the toxin. The water becomes contaminated underground by rocks that release the arsenic.

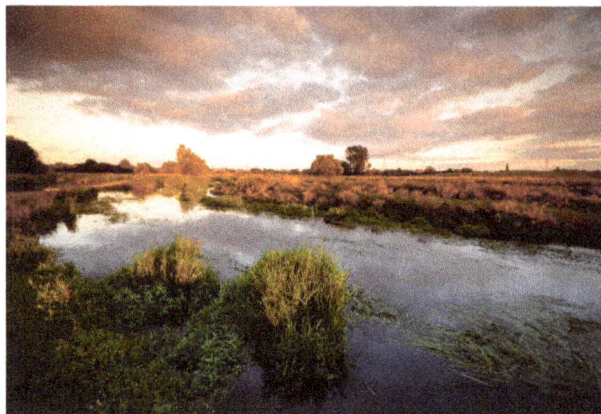

Groundwater possesses trace amounts of arsenic. On occasion,
these levels may exceed the amount a human can safely ingest.

Medical News Today (MNT) asked Dr. Daniel E. Brooks MD, Medical Director of the Banner Poison and Drug Information Center (BPDIC) about the risk of poisoning from contact with arsenic-contaminated underground rock.

He said: "There is no risk from touching rocks that contain arsenic. Transient contact with arsenic-containing rocks will not lead to effect absorption or clinical concerns for arsenic poisoning."

The World Health Organization (WHO) estimates that more than 200 million people worldwide are exposed to water that contains potentially unsafe levels of arsenic.

If proper safety measures are not taken, workers in certain industries may face a higher risk of toxicity.

These industries include:

- Glass production

- Smelting

- Wood treatment

- The production and use of some pesticides

The method through which arsenic enters the human body in these industries depends on the way the arsenic is being used.

For example, arsenic may be inhaled in the smelting industry, as there is inorganic arsenic in coke emissions. In the wood treatment industry, it may be absorbed through the skin if chemical containing arsenic makes contact.

There may be traces of arsenic in some foods, such as meat, poultry, and fish. Normally, poultry contains the highest level of arsenic, due to antibiotics in the chicken feed. Rice has also been found to potentially contain higher levels of arsenic than water.

Diagnosis

Pathological testing can confirm an instance of arsenic poisoning.

In areas and occupations with a risk of arsenic poisoning, it is important to monitor the levels of arsenic in the people at risk. This can be assessed through blood, hair, urine, and fingernail samples.

Urine tests should be carried out within 1 to 2 days of the initial exposure for an accurate measure of when the poisoning occurred. These tests can also be used to help diagnose cases of apparent arsenic poisoning.

Tests on hair and fingernails can determine the level of arsenic exposure over a period of up to 12 months. These tests can give an accurate indication of arsenic exposure levels, but they do not show what effects they may have on the person's health.

Treatment

The treatment depends on the type and stage of the arsenic poisoning.

Some methods remove arsenic from the human body before it causes any damage. Others repair or minimize the damage that has already occurred.

Treatment methods include:

- Removing clothes that could be contaminated with arsenic
- Thoroughly washing and rinsing affected skin
- Blood transfusions
- Taking heart medication in cases where the heart starts failing
- Using mineral supplements that lower the risk of potentially fatal heart rhythm problems
- Observing kidney function

Bowel irrigation is another option, a special solution is passed through the gastrointestinal tract, flushing out the contents. The irrigation removes traces of arsenic and prevents it from being absorbed into the gut.

Chelation therapy may also be used, this treatment uses certain chemicals, including dimercaptosuccinic acid and dimercaprol, to isolate the arsenic from the blood proteins.

Carbon Monoxide Poisoning

Carbon monoxide poisoning occurs when carbon monoxide builds up in your bloodstream. When too much carbon monoxide is in the air, your body replaces the oxygen in your red blood cells with carbon monoxide. This can lead to serious tissue damage, or even death.

Carbon monoxide is a colorless, odorless, tasteless gas produced by burning gasoline, wood, propane, charcoal or other fuel. Improperly ventilated appliances and engines, particularly in a tightly sealed or enclosed space, may allow carbon monoxide to accumulate to dangerous levels.

If you think you or someone you're with may have carbon monoxide poisoning, get into fresh air and seek emergency medical care.

Symptoms

Signs and symptoms of carbon monoxide poisoning may include:

- Dull headache
- Weakness
- Dizziness
- Nausea or vomiting
- Shortness of breath
- Confusion
- Blurred vision
- Loss of consciousness

Carbon monoxide poisoning can be particularly dangerous for people who are sleeping or intoxicated. People may have irreversible brain damage or even die before anyone realizes there's a problem.

Causes

Carbon monoxide poisoning is caused by inhaling combustion fumes. When too much carbon monoxide is in the air you're breathing, your body replaces the oxygen in your red blood cells with carbon monoxide. This prevents oxygen from reaching your tissues and organs.

Various fuel-burning appliances and engines produce carbon monoxide. The amount of carbon monoxide produced by these sources usually isn't cause for concern. But if they're used in a closed or partially closed space — cooking with a charcoal grill indoors, for example — the carbon monoxide can build to dangerous levels.

Smoke inhalation during a fire also can cause carbon monoxide poisoning.

Risk Factors

Exposure to carbon monoxide may be particularly dangerous for:

- Unborn babies: Fetal blood cells take up carbon monoxide more readily than adult blood cells do. This makes unborn babies more susceptible to harm from carbon monoxide poisoning.

- Children: Young children take breaths more frequently than adults do, which may make them more susceptible to carbon monoxide poisoning.

- Older adults: Older people who experience carbon monoxide poisoning may be more likely to develop brain damage.

- People who have chronic heart disease: People with a history of anemia and breathing problems also are more likely to get sick from exposure to carbon monoxide.

- Those in whom carbon monoxide poisoning leads to unconsciousness. Loss of consciousness indicates more severe exposure.

Complications

Depending on the degree and length of exposure, carbon monoxide poisoning can cause:

- Permanent brain damage.

- Damage to your heart, possibly leading to life-threatening cardiac complications.

- Fetal death or miscarriage.

- Death.

Prevention

Simple precautions can help prevent carbon monoxide poisoning:

- Install carbon monoxide detectors: Put one in the hallway near each sleeping area in your house. Check the batteries every time you check your smoke detector batteries — at least twice a year. If the alarm sounds, leave the house and call 911 or the fire department. Carbon monoxide detectors are also available for motor homes and boats.

- Open the garage door before starting your car: Never leave your car running in your garage. Be particularly cautious if you have an attached garage. Leaving your car running in a space attached to the rest of your house is never safe, even with the garage door open.

- Use gas appliances as recommended: Never use a gas stove or oven to heat your home. Use portable gas camp stoves outdoors only. Use fuel-burning space heaters only when someone is awake to monitor them and doors or windows are open to provide fresh air. Don't run a generator in an enclosed space, such as the basement or garage.

- Keep your fuel-burning appliances and engines properly vented. These include:

  o   Space heaters

  o   Furnaces

  o   Charcoal grills

  o   Cooking ranges

  o   Water heaters

  o   Fireplaces

  o   Portable generators

  o   Wood-burning stoves

  o   Car and truck engines

Ask your utility company about yearly checkups for all gas appliances, including your furnace.

- If you have a fireplace, keep it in good repair. Clean your fireplace chimney and flue every year.

- Keep vents and chimneys unblocked during remodeling. Check that they aren't covered by tarps or debris.

- Make repairs before returning to the site of an incident. If carbon monoxide poisoning has occurred in your home, it's critical to find and repair the source of the carbon monoxide before you stay there again. Your local fire department or utility company may be able to help.

- Use caution when working with solvents in a closed area. Methylene chloride, a solvent commonly found in paint and varnish removers, can break down (metabolize) into carbon monoxide when inhaled. Exposure to methylene chloride can cause carbon monoxide poisoning.

When working with solvents at home, use them only outdoors or in well-ventilated areas. Carefully read the instructions and follow the safety precautions on the label.

Chlorine Gas Poisoning

Chlorine is a chemical that inhibits bacterial growth in water. It's used to disinfect swimming pools and drinking water and sanitize sewage and industrial waste. It's also an active ingredient in several cleaning products.

Chlorine poisoning can occur when you touch, swallow, or inhale chlorine. Chlorine reacts with water outside of the body and on mucosal surfaces inside your body — including the water in your digestive tract — causing hydrochloric acid and hypochlorous acid to form. Both of these substances can be extremely poisonous to humans.

You may be most familiar with chlorine that's used in pools. However, most incidents of chlorine poisoning result from ingesting household cleaners, not pool water.

A few common household products and substances containing chlorine include:

- Chlorine tablets used in swimming pools
- Swimming pool water
- Mild household cleaners
- Bleach products

Symptoms of Chlorine Poisoning

Chlorine poisoning can cause symptoms throughout your body. Respiratory symptoms include coughing, difficulty breathing, and fluid inside the lungs.

Digestive system symptoms include:

- Burning in the mouth
- Swelling of the throat
- Throat pain
- Stomach pain
- Vomiting
- Blood in the stools

Chlorine exposure can damage your circulatory system. Symptoms of this problem can include:

- Changes in the pH balance of your blood

- Low blood pressure

- Serious injury to the eyes, including blurry vision, burning, irritation, and in extreme cases vision loss

- Skin damage, resulting from tissue injury with burns and irritation

Diagnosing Chlorine Poisoning

Chlorine poisoning has been known to occur in individuals over the years, so diagnosing it usually isn't difficult. In some cases, children may consume cleaning products that contain chlorine. This may be more difficult to diagnose since children sometimes can't tell you what they're feeling. Take children who show signs of chlorine poisoning to a hospital or emergency room immediately.

Treating Chlorine Poisoning

Seek medical assistance immediately if you or your child comes into contact with chlorine. Don't try to induce vomiting unless instructed by poison control or a medical professional.

If you get chlorine on your skin, immediately wash it with soap and water. If you get it in your eyes, flush them with running water for at least 15 minutes — take out contact lenses first if present. Remove any clothes that were on the areas of the body exposed to chlorine.

If you accidentally swallow chlorine, drink milk or water immediately, unless you experience vomiting or convulsions.

If you inhale chlorine, seek fresh air as soon as possible. Going to the highest possible ground to seek fresh air is helpful because chlorine is heavier than air.

Medical professionals will want to know the following information to treat your chlorine poisoning more effectively:

- Age

- Weight

- Clinical condition

- Product consumed

- Amount consumed

- Length of exposure

Once you've been admitted to the emergency room, a healthcare provider will measure and monitor your vital signs. This includes your pulse, temperature, blood pressure, oxygenation, and breathing rate. Doctors may also give you one or more of the following to help ease symptoms and help your body deal with the chlorine:

- Activated charcoal

- Supportive medications

- Intravenous fluid

- Supplemental oxygen

You might require placement of a breathing tube into your airway for mechanical ventilation if you have trouble breathing. Doctors might use a special tool to view your throat and determine if you have serious burns in your airways or lungs. A nasogastric tube may need to be inserted into your stomach to empty its contents.

Medical staff may need to wash affected skin at hourly intervals. Surgical removal of affected skin may be necessary if it's severely damaged.

Toluene Toxicity

Toluene toxicity refers to the harmful effects caused by toluene on the body.

Metabolism in Humans

While a significant amount of toluene, 25%–40%, is exhaled unchanged via the lungs, a greater proportion is metabolized and excreted via other pathways. The primary route of toluene metabolism is by hydroxylation to benzyl alcohol by members of the cytochrome P450 (CYP) superfamily. There are five CYPs which are important in toluene metabolism, CYP1A2, CYP2B6, CYP2E1, CYP2C8, and CYP1A1. The first four seem to be involved in the hydroxylation of toluene to benzyl alcohol. CYP2E1 seems to be the primary enzyme in the hydroxylation of toluene, accounting for roughly 44% of toluene metabolism; however, there is a great deal of ethnic variability, in the Finnish population for example the primary enzyme is CYP2B6. CYP2E1 catalyzes the formation of benzyl alcohol and p-cresol, while CYP2B6 produces comparatively little p-cresol.

It is believed that in humans, benzyl alcohol is metabolized to benzaldehyde by CYP rather than alcohol dehydrogenase; however, this belief does not appear to be universal. Benzaldehyde is in turn metabolized to benzoic acid, primarily by mitochondrial aldehyde dehydrogenase-2 (ALDH-2), while only a small percentage is metabolized by cytosolic ALDH-1.

Benzoic acid is metabolized to either benzoyl glucuronide or hippuric acid. Benzoyl glucuronide is produced by the reaction of benzoic acid with glucuronic acid, which accounts for 10–20% of benzoic acid elimination. Hippuric acid is also known as benzoylglycine and is produced from benzoic acid in two steps: first benzoic acid is converted to benzoyl-CoA by the enzyme benzoyl-CoA synthase; then benzoyl-CoA is converted to hippuric acid by benzoyl-CoA:glycine N-acyltransferase Hippuric acid is the primary urinary metabolite of toluene.

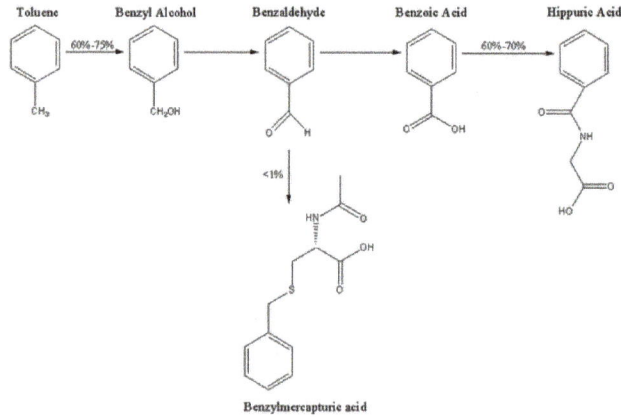

Ring hydroxylation to cresols is a minor pathway in the metabolism of toluene. The majority of the cresol is excreted unchanged in urine; however, some of the p-cresol and o-cresol is excreted as a conjugate. Studies in rats have shown that p-cresol is primarily conjugated with glucuronide to produce p-cresylglucuronide, though this may not be applicable to humans. o-cresol appears to be excreted mostly unchanged in urine or as the glucuronide or sulfate conjugate. There appears to be some dispute over whether m-cresol is produced as a metabolite of toluene or not.

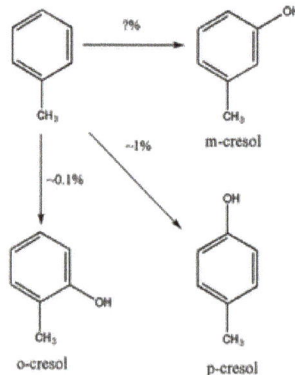

Environmental Influences

When exposure to toluene occurs there is usually simultaneous exposure to several other chemicals. Often toluene exposure occurs in conjunction with benzene and since they are to some degree metabolized by the same enzymes, the relative concentrations will determine their rate of elimination. Of course the longer it takes for toluene to be eliminated the more harm it is likely to do.

The smoking and drinking habits of those exposed to toluene will partially determine the elimination of toluene. Studies have shown that even a modest amount of acute ethanol consumption can significantly decrease the distribution or elimination of toluene from the blood resulting in increased tissue exposure. Other studies have shown that chronic ethanol consumption can enhance toluene metabolism via the induction of CYP2E1. Smoking has been shown to enhance the elimination rate of toluene from the body, perhaps as a result of enzyme induction.

The diet can also influence toluene elimination. Both a low-carbohydrate diet and fasting have been shown to induce CYP2E1 and as a result increase toluene metabolism. A low protein diet may decrease total CYP content and thereby reduce the elimination rate of the drug.

Measure of Exposure

Hippuric acid has long been used as an indicator of toluene exposure; however, there appears to be some doubt about its validity. There is significant endogenous hippuric acid production by humans; which shows inter- and intra-individual variation influenced by factors such as diet, medical treatment, alcohol consumption, etc. This suggests that hippuric acid may be an unreliable indicator of toluene exposure. It has been suggested that urinary hippuric acid, the traditional marker of toluene exposure is simply not sensitive enough to separate the exposed from the non-exposed. This has led to the investigation of other metabolites as markers for toluene exposure.

Urinary *o*-cresol may be more reliable for the biomonitoring of toluene exposure because, unlike hippuric acid, *o*-cresol is not found at detectable levels in unexposed subjects. o-Cresol may be a less sensitive marker of toluene exposure than hippuric acid. o-Cresol excretion may be an unreliable method for measuring toluene exposure because o-cresol makes up <1% of total toluene elimination.

Benzylmercapturic acid, a minor metabolite of toluene, is produced from benzaldehyde. In more recent years, studies have suggested the use of urinary benzylmercapturic acid as the best marker for toluene exposure, because: it is not detected in non-exposed subjects; it is more sensitive than hippuric acid at low concentrations; it is not affected by eating or drinking; it can detect toluene exposure down to approximately 15 ppm; and it shows a better quantitative relationship with toluene than hippuric acid or *o*-cresol.

Effects of Long-term Exposure

Serious adverse behavioral effects are often associated with chronic occupational exposure and toluene abuse related to the deliberate inhalation of solvents. Long-term toluene exposure is often associated with effects such as: psychoorganic syndrome; visual evoked potential (VEP) abnormality; toxic polyneuropathy, cerebellar, cognitive, and pyramidal dysfunctions; optic atrophy; and brain lesions.

The neurotoxic effects of long-term use (in particular repeated withdrawals) of toluene may cause postural tremors by upregulating GABA receptors within the cerebellar cortex. Treatment with GABA agonists such as benzodiazepines provide some relief from toluene-induced tremor and ataxia. An alternative to drug treatment is vim thalamotomy. The tremors associated with toluene misuse do not seem to be a transient symptom, but an irreversible and progressive symptom which continues after solvent abuse has been discontinued.

There is some evidence that low-level toluene exposure may cause disruption in the differentiation of astrocyte precursor cells. This does not appear to be a major hazard to adults; however, exposure of pregnant women to toluene during critical stages of fetal development could cause serious disruption to neuronal development.

Cyanide Poisoning

Cyanide is one of the most famous poisons — from spy novels to murder mysteries, it's developed a reputation for causing an almost immediate death.

But in real life, cyanide is a little more complicated. Cyanide can refer to any chemical that contains a carbon-nitrogen (CN) bond, and it can be found in some surprising places.

For example, it's found in many safe-to-eat plant foods, including almonds, lima beans, soy, and spinach.

You can also find cyanide in certain nitrile compounds used in medications like citalopram (Celexa) and cimetidine (Tagamet). Nitriles aren't as toxic because they don't easily release the carbon-nitrogen ion, which is what acts as a poison in the body.

Cyanide is even a byproduct of metabolism in the human body. It's exhaled in low amounts with every breath.

Deadly forms of cyanide include:

- Sodium cyanide (NaCN)

- Potassium cyanide (KCN)

- Hydrogen cyanide (HCN)

- Cyanogen chloride (CNCl)

These forms can appear as solids, liquids, or gases. You're most likely to encounter one of these forms during a building fire.

Symptoms of Cyanide Poisoning

Symptoms of toxic cyanide exposure may appear within a few seconds to several minutes after exposure.

You may experience:

- Overall weakness

- Nausea

- Confusion

- Headache

- Difficulty breathing

- Seizure

- Loss of consciousness

- Cardiac arrest

How severely you're affected by cyanide poisoning depends on:

- The dose

- The type of cyanide

- How long you were exposed

There are two different ways you can experience cyanide exposure. Acute cyanide poisoning has immediate, often life-threatening effects. Chronic cyanide poisoning results from exposure to smaller amounts over time.

Acute Cyanide Poisoning

Acute cyanide poisoning is relatively rare, and the majority of cases are from unintentional exposure.

When it does occur, symptoms are sudden and severe. You may experience:

- Difficulty breathing

- Seizure

- Loss of consciousness

- Cardiac arrest

If you suspect that you or a loved one is experiencing acute cyanide poisoning, seek immediate emergency medical attention. This condition is life-threatening.

Chronic Cyanide Poisoning

Chronic cyanide poisoning can occur if you're exposed to 20 to 40 parts per million (ppm) of hydrogen cyanide gas over a substantial period of time.

Symptoms are often gradual and increase in severity as time goes on.

Early symptoms may include:

- Headache
- Drowsiness
- Nausea
- Vomiting
- Vertigo
- Bright red flush

Additional symptoms may include:

- Dilated pupils
- Clammy skin
- Slower, shallower breaths
- Weaker, more rapid pulse
- Convulsions

If the condition remains undiagnosed and untreated, it can lead to:

- Slow, irregular heart rate
- Reduced body temperature
- Blue lips, face, and extremities
- Coma
- Death

Causes of Cyanide Poisoning

Cyanide poisoning is rare. When it does occur, it's typically the result of smoke inhalation or accidental poisoning when working with or around cyanide.

You may be at risk for accidental exposure if you work in certain fields. Many inorganic cyanide salts are used in the following industries:

- Metallurgy

- Plastic manufacturing
- Fumigation
- Photography

Chemists may also be at risk, as potassium and sodium cyanides are common reagents used in labs.

You may also be at risk for cyanide poisoning if you:

- Use excessive amounts of nail polish remover containing organic cyanide compounds like acetonitrile (methyl cyanide)
- Ingest excessive amounts of certain plant-based foods, such as apricot kernels, cherry rocks, and peach pits

Diagnosis of Cyanide Poisoning

If you're experiencing symptoms of acute cyanide poisoning, seek immediate emergency medical attention.

If you're experiencing symptoms of chronic cyanide poisoning, see your doctor right away. After discussing your symptoms, your doctor will perform a physical exam.

They'll also conduct blood tests to assess your:

- Methemoglobin level. Methemoglobin is measured when there is concern for smoke inhalation injury.
- Blood carbon monoxide concentration (carboxyhemoglobin level). Your blood carbon monoxide concentration can indicate how much smoke inhalation has occurred.
- Plasma or blood lactate level. Cyanide blood concentrations usually aren't available in time to help diagnose and treat acute cyanide poisoning, but they can offer later confirmation of poisoning.

Treatment Options

The first step to treating a suspected case of cyanide poisoning is to identify the source of exposure. This will help your doctor or other healthcare provider determine the appropriate decontamination method.

In the case of a fire or other emergency incident, rescue personnel will use protective gear like face masks, eye shields, and double gloves to enter the area and take you to a safe location.

If you have ingested cyanide, you may be given activated charcoal to help absorb the toxin and safely clear it from your body.

Cyanide exposure can affect oxygen intake, so your doctor may administer 100 percent oxygen via a mask or endotracheal tube.

In severe cases, your doctor may administer one of two antidotes:

- Cyanide antidote kit

- Hydroxocobalamin (Cyanokit)

The cyanide antidote kit consists of three medications given together: amyl nitrite, sodium nitrite, and sodium thiosulfate. The amyl nitrite is given by inhalation for 15 to 30 seconds, while sodium nitrite is administered intravenously over three to five minutes. Intravenous sodium thiosulfate is administered for about 30 minutes.

Hydroxocobalamin will detoxify cyanide by binding with it to produce nontoxic vitamin B-12. This medication neutralizes cyanide at a slow enough rate to allow an enzyme called rhodanese to further detoxify cyanide in the liver.

Ethylene Glycol Poisoning

Ethylene glycol poisoning is poisoning from ethylene glycol (a clear, colorless, odorless liquid with a sweet taste) that can produce dramatic and dangerous toxicity.

Ethylene glycol is found most commonly in antifreeze, automotive cooling systems, and hydraulic brake fluid. In an industrial setting it is also used as a solvent and in a variety of processes.

Many cases of ethylene glycol poisoning are due to the accidental ingestion of it by children. They may take in large amounts since the substance tastes good. Alcoholics may also drink it as a substitute for alcohol (ethanol).

Ethylene glycol is itself relatively nontoxic. However, it is metabolized (changed) in the body by the enzyme alcohol dehydrogenase into glycolic acid, glyoxylic acid and oxalic acid, which are highly toxic compounds.

Renal failure, acidosis and hypocalcemia may follow the intake of ethylene glycol. There can be widespread tissue injury in the kidney, brain, liver, and blood vessels. The result can be fatal.

The traditional treatment of ethylene glycol poisoning has been ethanol which competes for the attention of the enzyme (as a competitive substrate of alcohol dehydrogenase) and hemodialysis which removes ethylene glycol and its toxic metabolites from the blood. A new alcohol dehydrogenase inhibitor, fomepizole, was approved in 1997 for the treatment of ethylene glycol poisoning in patients at least 12 years old. It has also been used successfully with younger children.

Fluoride Toxicity

Fluoride toxicity is a condition in which there are elevated levels of the fluoride ion in the body. Although fluoride is safe for dental health at low concentrations, sustained consumption of large amounts of soluble fluoride salts is dangerous. Referring to a common salt of fluoride, sodium fluoride (NaF), the lethal dose for most adult humans is estimated at 5 to 10 g (which is equivalent to 32 to 64 mg/kg elemental fluoride/kg body weight). Ingestion of fluoride can produce gastrointestinal discomfort at doses at least 15 to 20 times lower (0.2–0.3 mg/kg or 10 to 15 mg for a 50 kg person) than lethal doses. Although it is helpful for dental health in low dosage, chronic exposure to fluoride in large amounts interferes with bone formation. In this way, the most widespread examples of fluoride poisoning arise from consumption of ground water that is abnormally fluoride-rich.

Recommended Levels

For optimal dental health, the World Health Organization recommends a level of fluoride from 0.5 to 1.0 mg/L (milligrams per liter), depending on climate. Fluorosis becomes possible above this recommended dosage. As of 2015, the United States Health and Human Services Department recommends a maximum of 0.7 milligrams of fluoride per liter of water – updating and replacing the previous recommended range of 0.7 to 1.2 milligrams issued in 1962. The new recommended level is intended to reduce the occurrence of dental fluorosis while maintaining water fluoridation.

Toxicity

Chronic

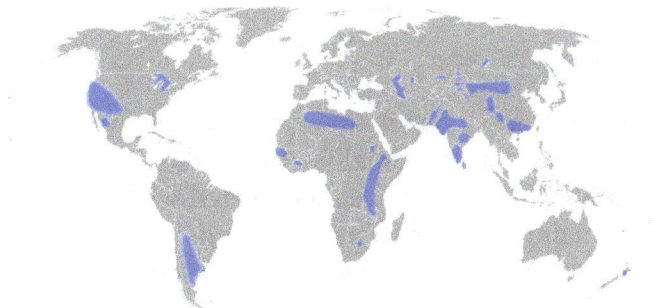

Geographical areas associated with groundwater having over 1.5 mg/L
of naturally occurring fluoride, which is above recommended levels.

In India an estimated 60 million people have been poisoned by well water contaminated by excessive fluoride, which is dissolved from the granite rocks. The effects are particularly evident in the bone deformities of children. Similar or larger problems are anticipated in other countries including China, Uzbekistan, and Ethiopia.

Acute

Historically, most cases of acute fluoride toxicity have followed accidental ingestion of sodium fluoride based insecticides or rodenticides. Currently, in advanced countries, most cases of fluoride exposure are due to the ingestion of dental fluoride products. Other sources include glass-etching or chrome-cleaning agents like ammonium bifluoride or hydrofluoric acid, industrial exposure to fluxes used to promote the flow of a molten metal on a solid surface, volcanic ejecta (for example, in cattle grazing after an 1845–1846 eruption of Hekla and the 1783–1784 flood basalt eruption of Laki), and metal cleaners. Malfunction of water fluoridation equipment has happened several times, including a notable incident in Alaska.

Occurrence

Organofluorine Compounds

Twenty percent of modern pharmaceuticals contain fluorine. These organofluorine compounds are not sources of fluoride poisoning. The carbon–fluorine bond is too strong to release fluoride.

Fluoride in Toothpaste

Children may experience gastrointestinal distress upon ingesting excessive amounts of flavored toothpaste. Between 1990 and 1994, over 628 people, mostly children, were treated after ingesting too much fluoride-containing toothpaste. "While the outcomes were generally not serious," gastrointestinal symptoms appear to be the most common problem reported. However given the low concentration of fluoride present in dental products, this is potentially due to consumption of other major components.

Fluoride in Drinking Water

Around one-third of the world's population drinks water from groundwater resources. Of this, about 10 percent, approximately 300 million people, obtains water from groundwater resources that are heavily contaminated with arsenic or fluoride. These trace elements derive mainly from leaching of minerals. Maps are available of locations of potential problematic wells via the Groundwater Assessment Platform GAP.

Effects

Excess fluoride consumption has been studied as a factor in the following:

Brain

Some research has suggested that high levels of fluoride exposure may adversely affect neurodevelopment in children, but the evidence is of insufficient quality to allow any firm conclusions to be drawn.

Bones

Whilst fluoridated water is associated with decreased levels of fractures in a population, toxic levels of fluoride have been associated with a weakening of bones and an increase in hip and wrist fractures. The U.S. National Research Council concludes that fractures with fluoride levels 1–4 mg/L, suggesting a dose-response relationship, but states that there is "suggestive but inadequate for drawing firm conclusions about the risk or safety of exposures at 2 mg/L".

Consumption of fluoride at levels beyond those used in fluoridated water for a long period of time causes skeletal fluorosis. In some areas, particularly the Asian subcontinent, skeletal fluorosis is endemic. It is known to cause irritable-bowel symptoms and joint pain. Early stages are not clinically obvious, and may be misdiagnosed as (sero-negative) rheumatoid arthritis or ankylosing spondylitis.

Kidney

Fluoride induced nephrotoxicity is kidney injury due to toxic levels of serum fluoride, commonly due to release of fluoride from fluorine-containing drugs, such as methoxyflurane.

Within the recommended dose, no effects are expected, but chronic ingestion in excess of 12 mg/day are expected to cause adverse effects, and an intake that high is possible when fluoride levels are around 4 mg/L. Those with impaired kidney function are more susceptible to adverse effects.

The kidney injury is characterised by failure to concentrate urine, leading to polyuria, and subsequent dehydration with hypernatremia and hyperosmolarity. Inorganic fluoride inhibits adenylate cyclase activity required for antidiuretic hormone effect on the distal convoluted tubule of the kidney. Fluoride also stimulates intrarenal vasodilation, leading to increased medullary blood flow, which interferes with the counter current mechanism in the kidney required for concentration of urine.

Fluoride induced nephrotoxicity is dose dependent, typically requiring serum fluoride levels exceeding 50 micromoles per liter (about 1 ppm) to cause clinically significant renal dysfunction, which is likely when the dose of methoxyflurane exceeds 2.5 MAC hours. (Note: "MAC hour" is the multiple of the minimum alveolar concentration (MAC) of the anesthetic used times the number of hours the drug is administered, a measure of the dosage of inhaled anesthetics).

Elimination of fluoride depends on glomerular filtration rate. Thus, patients with renal insufficiency will maintain serum fluoride for longer period of time, leading to increased risk of fluoride induced nephrotoxicity.

Teeth

The only generally accepted adverse effect of fluoride at levels used for water fluoridation is dental fluorosis, which can alter the appearance of children's teeth during tooth development; this is mostly mild and usually only an aesthetic concern. Compared to unfluoridated water, fluoridation to 1 mg/L is estimated to cause fluorosis in one of every 6 people (range 4–21), and to cause fluorosis of aesthetic concern in one of every 22 people (range 13.6–∞).

Thyroid

Fluoride's suppressive effect on the thyroid is more severe when iodine is deficient, and fluoride is associated with lower levels of iodine. Thyroid effects in humans were associated with fluoride levels 0.05–0.13 mg/kg/day when iodine intake was adequate and 0.01–0.03 mg/kg/day when iodine intake was inadequate. Its mechanisms and effects on the endocrine system remain unclear.

Effects on Aquatic Organisms

Fluoride accumulates in the bone tissues of fish and in the exoskeleton of aquatic invertebrates. The mechanism of fluoride toxicity in aquatic organisms is believed to involve the action of fluoride ions as enzymatic poisons. In soft waters with low ionic content, invertebrates and fishes may suffer adverse effects from fluoride concentration as low as 0.5 mg/L. Negative effects are less in hard waters and seawaters, as the bioavailability of fluoride ions is reduced with increasing water hardness. Seawater contains fluoride at a concentration of 1.3 mg/L.

Organophosphate Poisoning

Organophosphates are chemicals in insecticide used extensively in agriculture. When people, such as agricultural workers, are exposed to large quantities of organophosphates, these chemicals can be harmful.

When a person develops an illness as a result of organophosphate exposure, it is known as organophosphate poisoning.

Fast Facts on Organophosphate Poisoning

- Nearly 25 million cases of unintentional pesticide poisoning occur in the agricultural industry across the world each year.

- Cases are most common in regions where workers do not use or do not have access to protective gear, such as suits or masks.

- Symptoms and complications vary but can include death.

Signs and Symptoms

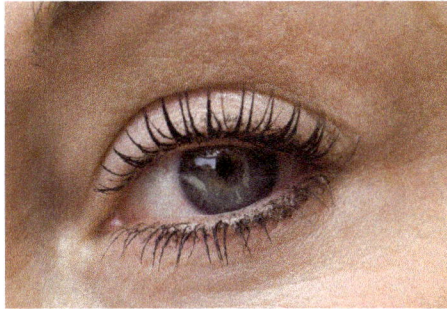

Possible symptoms of organophosphate poisoning include
narrowed pupils, and glazed over or watery eyes.

Organophosphate poisoning symptoms can range from mild to severe. In more severe cases, a person may die from the toxicity.

The length and strength of the exposure will determine the nature of someone's symptoms. Symptoms may start in as little as a few minutes or after several hours.

Symptoms of mild exposure to organophosphates include:

- Blurry or impaired vision
- Watery eyes
- Narrowed pupils
- Stinging eyes
- Nausea
- Runny nose
- Muscle twitching
- Glassy eyes
- Extra saliva
- Headache
- Muscle fatigue or weakness
- Agitation

Symptoms of moderate exposure to organophosphate include:

- Dizziness
- Very narrow pupils
- Fatigue

- Muscle tremors
- Muscle twitching
- Drooling
- Disorientation
- Wheezing or coughing
- Severe diarrhea
- Difficulty breathing
- Sneezing
- Uncontrolled urination or bowel movements
- Excessive phlegm
- Muscle weakness
- Severe vomiting

Symptoms of emergency-level exposure to organophosphate include:

- Confusion
- Narrow pupils
- Convulsions
- Coma
- Agitation
- Excessive secretions, such as saliva, sweat, tears, and mucus
- Irregular or slow heartbeat
- Collapsing
- Breathing that is ineffective stops

In addition to immediate signs and symptoms, organophosphate exposure can cause a number of long-term complications. Again, the severity of the complications depends on the extent and length of exposure.

Potential complications include:

- Paralysis
- Fertility issues
- Cancer

- Metabolic disorders, such as high blood sugar levels

- Inflammation of the pancreas

- Excess acid in the blood

- Brain and nerve problems

It is important to seek medical attention if a person shows signs of organophosphate poisoning to treat the potentially fatal condition effectively.

Causes and Risk Factors

Those who live or work near farmland may
be at risk of organophosphate poisoning.

People most at risk of organophosphate poisoning either live or work on or near a farm or farmland.

The most common way a person experiences unintended exposure is through direct contact with the skin or from breathing in the chemicals.

The pesticide is absorbed via the skin more rapidly if it is a liquid, oily, or if the skin is inflamed or has cuts or abrasions. It enters the lungs as a powder or through droplets in the air, including gas or vapors.

A less common method of exposure is through the consumption of water or food that is contaminated.

There have been a few reported cases of organophosphates used in terrorism. The reported cases occurred in Japan, where people were exposed to the chemical in an attempt to cause injury and death.

Finally, people may intentionally expose themselves to organophosphate. In these cases, a person often inhales or drinks some of the poison in an attempt to commit suicide.

Diagnosis

As with any poisoning through chemical exposure, a doctor will work with the person to figure out what chemical is causing the symptoms. The rapid onset of symptoms is how organophosphate exposure is often determined.

A doctor will also probably order blood work, and possibly urine samples if the person with organophosphate poisoning can cooperate. These tests will help determine how severe a person's exposure was and the right treatment.

Treatment

As with many poisoning and chemical exposure cases, the first step is stabilizing the person being treated. Emergency workers will often:

- Help the person return to normal breathing patterns.

- Decontaminate the person's body to prevent further damage.

- Use intravenous (IV) fluids to remove toxins from the blood and body.

In less severe cases, the person's breathing is often the priority. A doctor may still attempt to decontaminate the body, but the focus of treatment shifts to keeping a person breathing normally.

Atropine is a commonly used drug to help a person's breathing after chemical exposure. A doctor may also give medications, such as pralidoxime, to help with neuromuscular problems.

In the most severe cases where a seizure is likely, a doctor may prescribe benzodiazepines.

People who often work with organophosphates should discuss with their doctor the options for having an emergency injection of atropine.

Sulfuric Acid Poisoning

Sulfuric acid poisoning refers to ingestion of sulfuric acid, found in lead-acid batteries and some metal cleaners, pool cleaners, drain cleaners and anti-rust products.

Presentation

- Brown to black streak from angle of mouth

- Brown to black vomitus

- Brown to black stomach wall

- Black swollen tongue

- White (chalky white) teeth

- Blotting paper appearance of stomach mucosa

- Ulceration of esophagus (fibrosis and stricture)

- Perforation of stomach

- The stomach resembles a black spongy mass on post mortem

Treatment

For superficial injuries, washing (therapeutic irrigation) is important. Emergency treatments include protecting the airway, which might involve a tracheostomy. Further treatment will vary depending on the severity, but might include investigations to determine the extent of damage (bronchoscopy for the airways and endoscopy for the gastointestinal tract), followed by treatments including surgery (to debride and repair) and intravenous fluids.

Gastric lavage is contraindicated in corrosive acid poisoning like sulfuric acid poisoning. Bicarbonate is also contraindicated as it liberates carbon dioxide which can cause gastric dilatation leading to rupture of stomach.

Derived No-effect Level

Derived No-effect Level (DNEL) is defined as the level of chemical exposure above which humans should not be exposed. In human health risk assessment, the exposure level of each human population known to be or likely to be exposed is compared with the appropriate derived no-effect level (DNEL). The risk of a chemical substance to humans can be considered to be controlled/acceptable if the exposure levels estimated do not exceed the appropriate DNEL (i.e, exposure estimate/DNEL<1). In EU, DNELs shall be communicated to downstream users via extended safety data sheets (SDS) if they are available.

Derived No-effect Level (DNEL) Example - Methanol

The DNEL (oral route, systematic effects) of methanol for general population is 0.088 g/kg bw/day. It means that if an adult person (assuming 60kg body weight) intakes less than 5.28g (0.088*60) methanol per day by oral route, the risk of methanol causing ocular toxicity/vision damage can be controlled.

Calculation of Derived No-effect Level (DNEL)

The DNELs are usually calculated by dividing toxicological dose descriptors by an assessment factor. Since dose descriptors are usually obtained from animal studies, an assessment factor is required to allow for extrapolation to real human exposure situations and take into account of uncertainties.

The picture below shows you the relationship between LD50, LOAEL, NOAEL and DNEL on a typical dose-response curve.

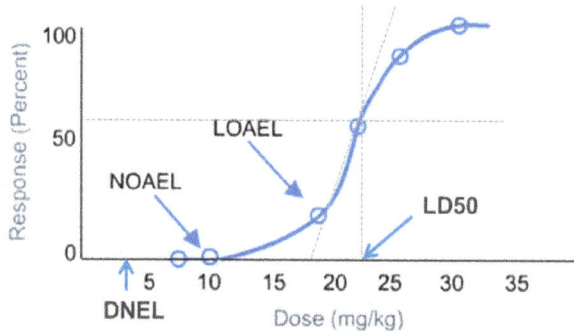

The table below is an example of how to divide NOAEL by assessment factors to get Derived No-effect Level (DNEL).

Effects	Adrenal Effects	Developmental Effects
NOAEL from animal study (oral route, long-term)	30 mg/kg bw/day (90d repeated dose)	80 mg/kg bw/day
Assessment Factor (Intraspecies)	10	10
Assessment Factor (Interspecies)	10	10
Assessment Factor (Duration)	3	1
Assessment Factor (Route extrapolation, data quality)	1	2
Total Assessment Factor (AF)	10x10x3x1=300	10x10x1x2=200
DNEL(oral route, long-term)	0.1mg/kg bw/day	0.4 mg/kg bw/day

In above case, the lowest DNEL (0.1mg/kg bw/day) will be used for risk assessment. If an adult (assuming weight is 60kg) intakes 12mg of a chemical substance per day, the estimated exposure (external dose per body weight) will be 0.2mg/kg bw/day. Since the exposure estimate is greater than DNEL, which will lead to a RCR>1, the risk will not be acceptable.

Number of Derived No-effect Level (DNELs) Required to Calculate

DNELs need to be derived for each relevant exposure pattern (*population, route, and duration of exposure*) and each relevant health effect (local and systemic effects). Under REACH, DNELs only need to be derived for chemical substances that are classified with health hazards according to CLP regulation/GHS.

The table below summarizes the number of DNELs you may need to derive:

Exposure Pattern	Workers	General Population
Acute – inhalation, systemic effects		
Acute – dermal, local effects		
Acute – inhalation, local effects		
Long-term – dermal, systemic effects		
Long-term – inhalation, systemic effects		
Long-term – oral, systemic effects	X(not relevant)	
Long-term – dermal, local effects		
Long-term – inhalation, local effects		

It is not always necessary to derive DNELs for all mentioned populations or routes or exposure duration. Depending on the exposure pattern and health effects, only relevant DNELs have to be derived. It may not always be possible to derive DNELs for all health effects. For many local effects (irritation), DNELs cannot be derived. This may also be the case, for example, for mutagen/carcinogen where no safe threshold level can be obtained. In these cases a semi-quantitative value, known as the DMEL or Derived Minimal Effect level may be developed.

To understand how many DNELs you actually have to derive, you need to understand the use of your substance and its exposure pattern (population, route, duration of exposure) as well as the type of its adverse health effects (local or systemic effects).

Populations

DNELs may have to be derived for workers and the general population. The general population includes consumers, and humans exposed via the environment.

Only DNELs for the relevant populations will have to be derived. For example, if a substance is only used as intermediate in a chemical plant and there is no exposure to general population, there is no need to derive DNEL for the general population.

Route of Exposure

In view of the likely exposure routes, DNELs may need to be derived for oral exposure (consumers/human via the environment), inhalation exposure (workers/consumers/human via the environment), and dermal exposure (workers/consumers, and potentially humans via the environment).

It is not always necessary to derive DNELs for all mentioned routes. Depending on the exposure pattern, only DNELs for the relevant routes of exposure will have to be derived. For example, there is no need to derive DNEL for inhalation route for a liquid substance which has very low vapor pressure and is unlikely to be inhaled. There is no need to derive DNELs for oral route for workers either.

Duration of Exposure

Depending on the exposure scenario, the exposure duration can vary from a single event to an exposure for several days/weeks/months per year, or it might even be continuous (as is, e.g., the case for humans exposed via the environment). Since the duration of exposure will often have an impact on the effect(s) that may arise, DNELs may have to be derived for various exposure duration.

Two main types of DNELs can be distinguished, DNEL *long-term* and DNEL *acute*:

- DNEL *long-term* is DNEL for effects that occur upon repeated exposure. It shall always be derived. Toxicity studies that give information on these possible 'long-term' effects of a substance are: repeated dose toxicity studies, reproductive toxicity studies (including developmental toxicity studies), and carcinogenicity studies. 'Long-term' is here used as a more general term, including, e.g., sub-chronic (usually 90 days) as well as chronic (usually 1.5 - 2 years) studies.

- DNEL *acute* can generally be defined as a DNEL for effects that occur after exposure for a short period of time (from minutes to a few hours). There is no adverse health effects from short exposure, there is no need to derive *DNEL acute*. The potential for short-term high level (i.e. peak) inhalation exposure is of most concern for workers, and hence, the occupational exposure assessment should always consider the possibility for such peak inhalation exposures.

Local Effect vs Systemic Effect

Depending on the substance and the type of adverse health effects, DNELs may have to be established for systemic effects, for local effects or for both:

- A *local effect* is an effect that is observed at the site of first contact, caused irrespective of whether a substance is systemically available. This mainly includes skin/eye irritation/corrosion, skin or respiratory sensitization.

- A *systemic effect* is defined as an effect that is normally observed distant from the site of first contact, i.e., after having passed through a physiological barrier (mucous membrane of the gastrointestinal tract or of the respiratory tract, or the skin) and becomes systemically available. This includes acute toxicity, repeated dose toxicity, genotoxicity, reproductive toxicity and carcinogenicity.

For systemic effects, the units of DNELs are mg/m3 for inhalation, and mg/kg bw ormg/kg bw/day for oral and dermal exposure.

For local effects, the units of DNELs are mg/m3 for inhalation; and for dermal exposure: mg/cm2 skin, mg/person/day (e.g., calculated based on the deposited amount per cm2 times the actually exposed body area), or a measure of concentration (% or ppm).

References

- Gosselin, RE; Smith RP; Hodge HC (1984). Clinical toxicology of commercial products. Baltimore (MD): Williams & Wilkins. pp. III-185–93. ISBN 0-683-03632-7

- Chlorine-poisoning-symptoms, health: healthline.com, Retrieved 16 April 2018

- Joseph A. Cotruvo. "Desalination Guidelines Development for Drinking Water: Background" (PDF). Retrieved January 26, 2015

- Hjelm, EW; Näslund PH; Wallén M (1988). "Influence of cigarette smoking on the toxicokinetics of toluene in humans". Journal of Toxicology and Environmental Health. 25(2): 155–63. doi:10.1080/15287398809531197. PMID 3172270

- Cyanide-poisoning, health: healthline.com, Retrieved 10 June 2018

- Jay D. Shulman; Linda M. Wells (1997). "Acute Fluoride Toxicity from Ingesting Home-use Dental Products in Children, Birth to 6 Years of Age". Journal of Public Health Dentistry. 57(3): 150–158. doi:10.1111/j.1752-7325.1997.tb02966.x. PMID 9383753

- How-to-Derive-Derived-No-Effect-Level-(DNEL): chemsafetypro.com, Retrieved 18 March 2018

- Inoue, O; Seiji K; Watanabe T; Chen Z; Huang MY; Xu XP; Qiao X; Ikeda M (May 1994). "Effects of smoking and drinking habits on urinary o-cresol excretion after occupational exposure to toluene vapor among Chinese workers". American Journal of Industrial Medicine. 25 (5): 697–708. doi:10.1002/ajim.4700250509. PMID 8030640

Chapter 7
Organ Specific Toxic Effects

Specific-target organ toxins damage specific organs only. Toxins can target the respiratory system, reproduction system, neurological system, cardiovascular system, etc. The various toxic effects of organ specific toxins have been discussed in this chapter.

Cardiotoxicity

Cardiotoxicity is also Known as Cardiac toxicity. Cardiac toxicity is damage to the heart by harmful chemicals. As part of your treatment, you may be given toxins (drugs) to kill cancer cells. A side effect is that the normal cells in and around your heart can also be killed. Besides cell death, other types of cardiac toxicity from cancer treatment include:

- Cardiomyopathy is when heart muscle becomes weakened, enlarged, thickened, or stiff. This can lead to changes in heart rhythm or to heart failure.

- Myocarditis is inflammation (swelling) of the heart. This can lead to changes in heart rhythm or heart failure.

- Pericarditis is inflammation (swelling) of the sac surrounding the heart. This can cause chest pain or heart failure.

- Acute coronary syndromes are caused by blood vessel damage, which reduces blood flow to the heart. This can cause chest pain or a heart attack (myocardial infarction).

- Congestive heart failure is when the heart is unable to pump enough blood throughout the body. Chemotherapy can cause mild or severe damage to the heart. If severe, congestive heart failure or other life-threatening problems can occur. A heart transplant may even be needed.

Cancer Treatment that Causes Cardiotoxicity

Chemotherapy

The chemotherapy drugs that most commonly cause heart damage are anthracyclines. Anthracyclines are used to treat many types of cancers. Examples include some types of leukemia, lymphoma, sarcomas as well as bladder, bone, breast, head and neck, kidney, skin, stomach, and other cancers. Anthracyclines include:

- Daunorubicin (Cerubidine),

- Doxorubicin,

- Doxorubicin liposome injection (Doxil),

- Epirubicin (Ellence),

- Idarubicin (Idamycin PFS),

- Valrubicin (Valstar)

In rare cases, cyclophosphamide can damage the heart. This chemotherapy drug is an alkylating agent. It is sometimes used to treat some types of leukemia, lymphoma, myeloma, and sarcomas as well as bone, breast, central nervous system, ovarian, and skin cancers.

Targeted Therapy

Targeted therapy drugs that can cause damage to the heart include trastuzumab (Herceptin), bevacizumab (Avastin), lapatinib (Tykerb), sunitinib (Sutent), and sorafenib (Nexavar). These drugs are used in a variety of cancer types.

Radiation Therapy

Many patients with cancer get radiation therapy to the chest. Examples of cancers that may be treated with radiation include breast cancer, lung cancer, lymphoma, and some childhood cancers. This radiation can damage the blood vessels that bring blood to the heart. Patients who are treated with both anthracyclines and radiation to the chest are at high risk for heart problems. So are patients who are treated with both anthracyclines and trastuzumab.

People who are at the Increased Risk of Cardiotoxicity

Patients who are older, young children, and women have a greater risk for cardiac toxicity. In addition, patients who have other health conditions at the same time as cancer are at increased risk. This is especially true for patients with signs of heart trouble before cancer.

Symptoms of Cardiotoxicity

You may not notice any symptoms. Then again, you might have chest pain or notice changes in your heart rhythm. Arrhythmias are health problems with the speed or rhythm of the heartbeat. Atrial fibrillation is the most common type of arrhythmia. It is an abnormal beating of the heart's upper left chamber. It is common not to notice atrial fibrillation.

Your doctor may tell you that you have a decline in LVEF (left ventricular ejection fraction). This means your heart isn't pumping as much blood with each heartbeat as it should. Your treatment will likely be changed if this happens, and you may be started on drugs to help your heart.

If the damage is severe, you may have congestive heart failure. You will probably feel very tired and have trouble breathing. You will first notice the shortness of breath when you are active. Later, you will be short of breath even while resting. Congestive heart failure causes you to gain weight and your ankles to swell. You may also find it uncomfortable to lie on your back.

You might not notice these symptoms until many months or even years after you have completed cancer treatment. Tell your doctor if you any of these symptoms start.

Diagnosis of Cardiotoxicity

Your heart will likely be checked before you start treatment. This will tell your doctor about your baseline (pre-treatment) heart health. Your heart will also be checked regularly during treatment, when starting different treatments, and after you are done with treatment. These are some common tests used to check your heart:

- Physical exam is a review of your body for signs of disease. During this exam, your doctor may listen to your heart with a stethoscope. If it does not sound normal, there may be damage to your heart.

- Chest X-ray is a type of imaging test that makes pictures of the insides of your chest. Your doctor can see if your heart looks too big or if fluid is building up in your lungs.

- Echocardiogram is a type of imaging test that uses ultrasound. Ultrasound bounces sound waves off of organs to make pictures. Your doctor may use an echocardiogram to see your heart in action. Your doctor will be able to see if your heart is pumping enough blood. This is the test used to measure LVEF.

- ECG (electrocardiogram) is a test that measures the electrical activity of the heart. It lets your doctor see your heart rhythm in detail.

- MUGA (multi-gated acquisition) scan is a test that injects a radiotracer into your vein. The radiotracer attaches to red blood cells. Next, a special camera is used to see the radiotracer go through your heart. Your doctor can then see how well your heart is pumping. Your doctor can also see how well the blood vessels to the heart are working.

- Troponin blood tests are done less often. Troponins are proteins found in heart muscle. Troponins are released by dying heart cells and then enter the bloodstream. Troponins may be in your blood even before a decline in LVEF is seen. The use of troponins in the blood to predict heart problems is still being studied.

Prevention of Cardiotoxicity

Cardiac toxicity can often be prevented by giving less of the cancer drug. Also, you could get lower doses more often instead of larger doses less often. There may be drugs that are less toxic. For example, doxorubicin liposome injection, a liposomal anthracycline, is less toxic than regular doxorubicin.

The problem is balance. You don't want to have more treatment than is needed to treat the disease. Then again, you don't want to undertreat the cancer just to lower the chance of side effects. You should talk to your doctor about the balance of risks and benefits of cardiotoxic treatment for you.

For breast cancer, sometimes dexrazoxane (Zinecard) is used to prevent or lessen heart damage caused by doxorubicin or epirubicin. It is given right before either drug to protect your heart.

Other drugs are now being tested for prevention of damage to the heart in high-risk patients. These include enalapril, an ACE (**a**ngiotensin-**c**onverting **e**nzyme) inhibitor, and carvedilol, a beta-blocker.

Treatment of Cardiotoxicity

If treatment for cardiac toxicity is started early, it is more likely that serious heart damage will be prevented. If you have heart failure because of cancer treatment, you will be treated like other patients with heart failure. You may be given some combination of these medicines:

- A diuretic (to control your fluids),

- An ACE inhibitor (to control your blood pressure),

- A beta-blocker (to control your blood pressure),

- A digitalis drug (to make your heart stronger and to regulate heart rhythm).

In severe cases, a heart transplant may be needed.

Dermal Toxicity

Dermal toxicity is the ability of a substance or poison to adversely affect the skin. Defatting of skin, rashes, and irritation are common with various toxins, drugs, ketones, chlorinated compounds, alcohols, nickel, phenol, trichloroethylene, zinc deficiency, etc. (Acneiform eruptions, or folliculitis, often begin as facial erythema that progresses to papules and pustules and spreads to the upper trunk. Causes of folliculitis in cancer patients include actinomycin-D (Cosmegen)—the most common—as well as epidermal growth factor receptor inhibiting agents, such as gefitinib (Iressa) and cetuximab (Erbitux)).

Dermal toxicity is evaluated by applying the chemical to the skin for 6 h a day for 21 days in rat studies, or 28 days in rabbit studies. Feeding studies are used to evaluate the toxicological effects of the chemical when a known dose is administered orally.

The oral feeding studies involve giving rats, mice, or dogs diets containing the chemical for various lengths of time. Rat and mouse feeding studies are conducted for intervals of 28 days, 90 days, 1 year, and for the lifetime of the animals (24 months for rats, 18 months for mice). When dogs are used as the test animal, studies are usually conducted for 28 days, 90 days, 1 year, or 2 years. In all cases, animals are divided into test groups of 10–50 rats or mice and four to six dogs. At least four test groups are used in each study, one control group receiving no chemical and three groups receiving low, medium, or high concentrations of the chemical in their diets. In these studies, urinalysis, hematology, and clinical chemistry parameters are evaluated and gross and microscopic pathological examinations are performed on up to 50 tissue samples. Maximally tolerated doses are tested in order to demonstrate toxicity (up to 1000 mg/kg/day in the diet). In this fashion, it is possible to determine whether a chemical damages or alters any organ or tissue. In addition, it is possible to establish levels of the chemical that produce the NOEL, and the lowest level at which effects are noted (LOEL). The response of repeated exposure of rats and dogs to the selected triazine herbicides are presented in table.

With the exception of rats and dogs fed terbuthylazine and cyanazine and rats fed metribuzin, the NOEL values were 2.5 mg/kg/day or higher, and LOEL values were 15 mg/kg/day or higher. Furthermore, the most common observation was not a specific organ or tissue effect, but a reduction in body weight gain. Microscopically, the liver was the most common target organ.

Acute dermal toxicity is another common component of the testing battery, again the OECD protocol (Organization for Economic Cooperation and Development 2008) or a variation employing a limit test or the FHSA, TSCA, or DOT protocols are used. Although some details of the tests may vary, the basic outline of these tests consist of a 24 h exposure of test material under an occluded or semioccluded bandage to the shaved skin of animals (rabbits are commonly used, although rats can be when a test article may be in limited supply). As for the oral protocols, animals are observed at least daily over a period of 14 days for clinical signs of toxicity, with body weights recorded prior to dosing and weekly thereafter until termination. At termination (either scheduled day 14 or due to moribund condition of the animals), all animals receive a gross necropsy with histologic examination of any grossly observed abnormal organs. Again, a limit test of 2000–5000 mg kg−1 is allowed but, failing this, at least three dose levels are generally required to establish a median lethal dose for the compound. Vehicle controls are not included unless the toxicity of the vehicle is unknown.

Ototoxicity

Ototoxicity refers to drug or chemical-related damage to the inner ear, resulting in damage to the organs responsible for hearing and balance. Such damage can lead to temporary or permanent hearing loss, and loss of balance. Ototoxic substances include several therapeutic medicines (e.g. aminoglycoside antibiotics, some anti-cancer agents, loop diuretics, anti-malarial drugs and aspirin), and other environmental substances (e.g. mercury, lead and arsenic). The table below lists some of the most commonly used ototoxic medications and substances.

Table: Common substances known to be associated with ototoxicity

Type/group	Name of oxotoxic substance
Aminoglycoside antibiotics	Gentamicin, streptomycin, tobramycin, neomycin, netilimicin, kanamycin, amkicacin, dihydrostreptomycin, ribostamycin
Non-aminoglycoside antibiotics	Vancomycin, erythromycin
Loop diuretics	Furosemide, ethacrynic acid, bumetanide, torsemide
Chemotherapeutic agents	Cisplatin, carboplatin, nitrogen mustard
Salicylates	Aspirin
Anti-malarial drugs	Quinine, chloroquine
Environmental chemicals and other substances	Lead, mercury, carbon monoxide, arsenic, carbon disulfide, tin, hexane, toluene, alchohol

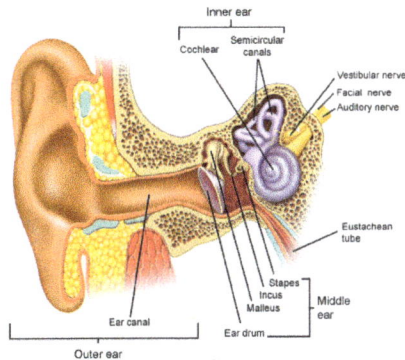

Statistics

Ototoxicity can affect anyone receiving treatment from the medications listed above. However, the likelihood of developing ototoxicity is highly dependent on which drug is being administered, duration of use, and several other underlying factors, such as kidney function and the use of other ototoxic drugs at the same time.

Around 10% of people taking aminoglycoside antibiotics experience ototoxicity, although up to 33% has also been reported in adult patients, with a 3% chance of that damage being permanent. Generally, ototoxicity occurs more commonly in adults than in children and babies, where the incidence rate is about 2%. Vestibular toxicity from aminoglycosides is documented to occur in as many as 4% of adult patients, and around 2% of patients treated with gentamicin report hearing loss.

In patients being treated with the anti-cancer agent cisplatin, ototoxicity can occur in as many as 50% of patients. The rate and severity of ototoxicity is higher in children, and in patients who have had previous radiation therapy to the head and neck. Approximately 60% of children receiving platinum-based chemotherapy experience hearing loss.

There is a 6% and 0.7% incidence of ototoxicity from the diuretics, furosemide and ethacrynic acid, respectively. Up to 1% of patients report ototoxicity from aspirin, though this occurs most commonly in the elderly.

Risk Factors

There are certain factors that may put patients at increased risk for ototoxicity:

- Dose and duration of therapy
- Infusion rate and cumulative lifetime dose
- Impaired kidney function, which can lead to rapid accumulation of the ototoxic drug
- Concurrent administration of another ototoxic drug (e.g. aminoglycosides and loop diuretics)
- Age
- Pre-existing hearing loss, sensorineural hearing loss
- Exposure during pregnancy
- Previous exposure to head and neck radiation (for chemotherapeutic agents)
- Genetic susceptibility
- Family history of ototoxicity

Progression

Symptoms of ototoxicity may occur rapidly or appear months following administration. Often in the early stages, ototoxicity goes undiagnosed, such as when hearing loss is very minimal or restricted to high pitched sounds. It is usually when hearing loss reaches the lower speech frequencies that patients notice anything, and by that stage permanent damage has already occurred.

Aminoglycosides are known for their potential to cause permanent hearing loss, normally preceded by high pitched tinnitus and a gradual loss of hearing that begins in the higher frequencies. Loss of vestibular sensitivity can also be permanent, and can cause the patient's vision to oscillate or bounce.

Symptoms of aspirin and quinine toxicity are dependent on dose and are usually reversible. They are characterised by tinnitus and mild hearing loss. Large doses of quinine, however, have been known to cause permanent hearing loss, particularly in elderly patients taking long-term medication for leg cramps.

Chemotherapeutic agents are well known causes of hearing loss that can be severe and permanent. Hearing loss usually begins as a loss of high frequencies in both ears, and progresses to a loss of all frequencies. Hearing loss presents as a sensation of hearing muffled voices. Vestibular effects (e.g. loss of balance, uncoordination, vertigo) are also common. Hearing loss usually occurs after 1-2 weeks of treatment, although it can often be delayed for up to 6 months post treatment.

Loop diuretics can cause ringing in the ears or reduce hearing. This is usually reversible once treatment is stopped.

Symptoms

The most common symptoms experienced from ototoxicity are:

- Tinnitus or ringing in the ears
- Bilateral or unilateral hearing loss
- Dizziness
- Uncoordination in movements
- Unsteadiness of gait
- Oscillating or bouncing vision

Diagnosis

Several specific audiologic tests are available which your doctor may perform. They include various hearing and balance tests. These should be conducted prior to the beginning of treatment with a known ototoxic agent, as well as during treatment and after treatment has stopped.

- Pure tone air conduction test: Can detect very small changes even before onset of tinnitus, as most ototoxic agents produce hearing loss in the highest frequencies first. Early detection permits modification of treatment before speech frequencies are affected.
- Pure tone bone conduction: Used to determine sensorineural function.
- Word recognition tests.

- Romberg's test: Balance test to detect vestibular damage.

For infants and critically ill patients who are bed-bound or comatose, alternative tests are available:

- Otoacoustic emission (OAE): Involves the use of a microphone to measure signals that are produced by the cochlea.

- Auditory brainstem response (ABR): Measures auditory function that utilizes responses produced by the auditory nerve and the brainstem. Helps differentiate sensory from neural hearing loss.

Prognosis

How well a patient recovers from ototoxicity is dependent on the type of drug and the dose and duration of treatment. Usually hearing loss from cisplatin treatment is irreversible, whereas that occurring from salicylates and quinine is most often reversible.

Most environmental chemicals are associated with permanent hearing loss. Mercury has been associated with permanent balance problems.

If symptoms are detected early, then the chance of recovery is much higher. However, most patients do not notice any significant changes until it is too late.

Genotoxicity

Genotoxicity is a word used in genetics that describes the possession of substance that has destructive effect on the genetic material of the cell (DNA, RNA) thus affecting the integrity of the cell. Genotoxins are mutagens that can cause genotoxicity leading to the damage of DNA or chromosomal material thus causing mutation. Genotoxins can include chemical substance as well as radiation. Genetic toxicology is the branch of science that deals with the study of agents or substances that can damage the cell's DNA and chromosomes. It is noted that often genotoxicity is confused with mutagenicity. All mutagens are genotoxic however all genotoxic substances are not mutagenic.

Genotoxins can be of the following category depending on its effects:

- Carcinogens or cancer causing agents

- Mutagens or mutation causing agents

- Teratogens or birth defect causing agents

The damage of genetic material of somatic cells may lead to malignancy (cancer) in eukaryotic organisms. Whereas the genetic damage of the germ cells may lead to heritable mutations causing birth defects. Mutations can be of any form; which may include duplication, insertion or deletion of genetic information. These mutations can cause varying range of problems in the host, from a wide variety of diseases to cancer One of the best ways to control the damage due to mutagens and carcinogens is to identify the substance or chemical, i.e. antimutagens/anticlastogens (which suppress or inhibit the mutagenesis process by directly acting on the cell mechanism) and demutagens (which destroy or inactivate the mutagens partially or fully thereby affecting less population of cell) from the medicinal plants so that it can be used as antimutagenic and anticarcinogenic food or drug additives.

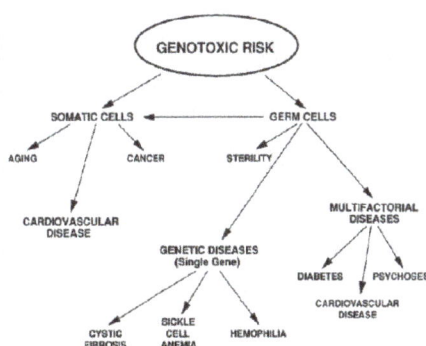

Figure: Risk of genotoxicity.

Importance of Genotoxicity Studies

Genotoxicity studies can be defined as various in-vitro and in-vivo tests designed to identify any substance or compounds which may induce damage to genetic material either directly or indirectly by various mechanisms. These tests should enable the identification of hazard with respect to DNA damage and fixation. Genetic change plays only a part in the complex process of heritable effects and malignancy which include the fixation of the damage to the DNA by gene mutation or large scale chromosomal damage or recombination or numerical chromosomal changes. These tests play an important role in predicting if the compounds have the potential to cause genotoxicity and carcinogenicity by testing them positive. As a part of safety evaluation process, regulatory authorities all over the globe require information on the genotoxic potential of the new drugs. Genotoxicity is usually evaluated along with other toxicological end points during the safety assessment.

During the early testing stages; the same testing assays are carried out for predicting both the potential heritable germ cell damage as well as the carcinogenicity because these endpoints have common precursors. The relationship between exposure to particular chemical and carcinogenesis has been established whereas such relationship has been difficult to establish for heritable diseases, genotoxic studies have been mainly associated and used for the prediction of carcinogenicity of a compound.

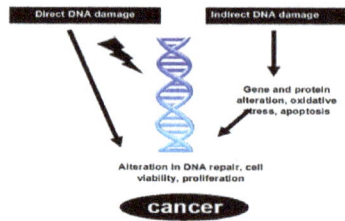

Figure: Relationship between genotoxicity and carcinogenicity.

Classifiaction of Carcinogens

EU Classification of Carcinogens

- Carcinogen category 1-shown to cause cancer in humans.

- Carcinogen category 2-causes cancer in animal tests, and most probably also in humans.

- Carcinogen category 3-possibly carcinogenic, but evidence supporting carcinogenicity is inadequate for the classification to category 2.

Iarc (International Agency for Research on Cancer) Classification of Carcinogens

- IARC class 1-The substance is carcinogenic to humans.

- IARC class 2A-The substance is probably carcinogenic to humans.

- IARC class 2B-The substance is possibly carcinogenic to humans.

- IARC class 3-The substance is not classifiable to as to its carcinogenicity to humans.

- IARC class 4-The substance is probably not carcinogenic to humans.

Agents that can Cause Direct or Indirect Damage to the DNA

Reactive oxygen species are known to be genotoxic in nature, thus any chemical or substance that may increase the reactive oxygen species (ROS) production might evidently add to the endogenously produced ROS and may lead to non-linear relationships of dose-effect. The following agents are capable of damaging the DNA directly or indirectly;

- Electrophilic species that form covalent adducts to the DNA

- Reactive oxygen species

- Ultra violet and ionizing radiations.

- Nucleoside analogues

- Topoisomerase inhibitors

- Protein synthesis inhibitors

- Some herbal plants like Aconite, Alfa-alfa, Calamus, Aloe vera, Isabghol etc

Anti-mutagen is any agent that decreases the effect of spontaneous and induced mutations. There are mainly two mechanisms of anti-mutagenesis:

a. Desmutagenesis in which the factors on the mutagens are somehow inactivated,

b. Bio-antimutagenesis, in which the factors act on the process of mutagenesis or by repairing the damaged DNA that result in the decreased frequency of DNA mutation. Our cells have several DNA repair system by which they try to control the DNA mutations naturally.

The five major pathways through which the cell repairs the damaged DNA are:

- Direct repair

- Base excision repair(BER)

- Nucleotide excision repair (NER)

- Mismatch repair

Single/double strand break repair.

Mechanism of Genotoxicity

The damage to the genetic material is caused by the interactions of the genotoxic substance with the DNA structure and sequence. These genotoxic substances interact at a specific location or base sequence of the DNA structure causing lesions, breakage, fusion, deletion, mis-segregation or nondisjunction leading to damage and mutation. For example, in its high-valent oxidation state the transition metal chromium interacts with the DNA so that DNA lesions occur leading to carcinogenesis. Researchers have found that the mechanism of damage and base oxidation products for the interaction between DNA and high-valent chromium are relevant to in-vivo formation of DNA damage leading to cancer in chromate- exposed human population, thus making high valent chromium a carcinogen.

Reactive oxygen species causes one of the most abundant oxidative lesions in DNA and is 8- hydroxydeoxyguanosine (8- OHdG), which is a potent mutagenic lesion. Oxidants as well as free radical when present in the cellular system can adversely affect and alter the structure of lipids, proteins as well as DNA. Reactive aldehydes like 4-hydroxynonenal (4-HNE) are generated by the decomposition of lipid peroxyl radicals or primary free radical intermediate of lipid peroxidation. 4-Hydroxynonenal is involved in many of the oxidative stress related diseases such as atherosclerosis, fibrosis, neurodegenerative diseases it. Many studies have indicated that 4-Hydroxynonenal can stimulate the cell proliferation, differentiation as well as cytoprotective response through its effects on various signalling pathway.

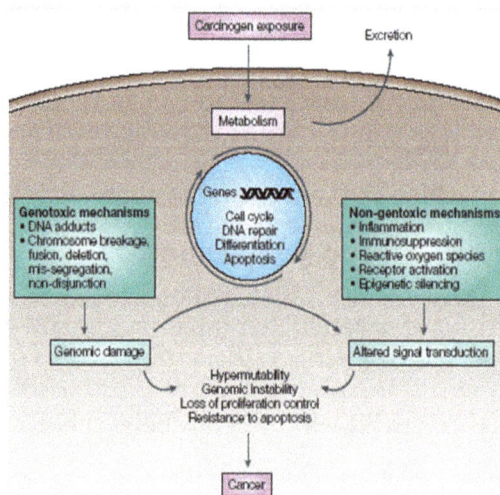

Figure: Mechanism of carcinogenesis.

Standard Test Battery for Genotoxicity

The standard test battery for genotoxicity recommends the following for genotoxicity.

Table: The standard test battery for genotoxicity recommends the following for genotoxicity Evaluation.

TG 471	Bacterial Reverse Mutation Test (Ames Test)
TG 472	Genetic Toxicology: Escherichia coli, reverse assay
TG 473	In-Vitro Mammalian Chromosome Aberration Test
TG 474	Mammalian Erythrocyte Micronucleus Test
TG 475	Mammalian Bone Marrow Chromosome Aberration Test
TG 476	In-Vitro Mammalian Cell Gene Mutation Test
TG 477	Genetic Toxicology: Sex-linked Recessive Lethal Test in Drosophila melanogaster
TG 478	Genetic Toxicology: Rodent Dominant Lethal Test
TG 479	Genetic Toxicology: In-Vitro Sister Chromatid Exchange Assay in Mammalian Cells
TG 480	Genetic Toxicology: Saccharomyces cerevisiae, Gene Mutation Assay
TG 481	Genetic Toxicology: Saccharomyces cerevisiae, Mitotic Recombination Assay
TG 482	Genetic Toxicology: DNA Damage and Repair, Unscheduled DNA Synthesis in Mammalian Cells In-Vitro
TG 483	Mammalian Spermatogonial Chromosome Aberration Test
TG 484	Genetic Toxicology: Mouse Spot Test
TG 485	Genetic Toxicology: Mouse Heritable Translocation Assay
TG 486	Unscheduled DNA Synthesis (UDS) Test with Mouse Liver Cells In-Vitro
TG 487	In-Vitro Mammalian Cell Micronucleus Test

- Testing for gene mutation in bacteria.

- In-vitro: cytogenetic evaluation of chromosomal damage with mammalian cells or mouse lymphoma assay.

- In-vivo: test for chromosomal damage using rodent hematopoietic cells.

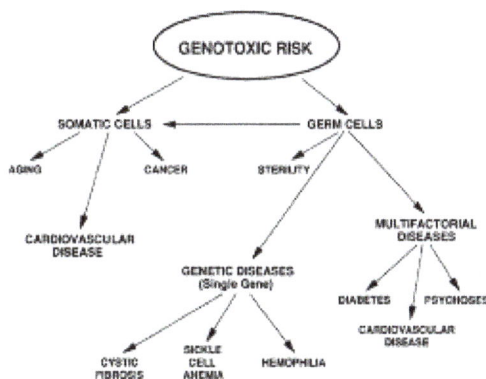

Figure: Flow chart on Genotoxicity Testing.

Purpose of Genotoxicity Assays

Assays even though inexpensive, have high statistical power and can be reproduced and have the ability to detect a wide variety of genotoxic end-points. It also allows the detection of a drug's potential to cause genotoxicity even in the early stage of drug development. They are designed in such a way that it can be more sensitive to damage so as to enhance the identification of hazard.

In-vitro Testing Methods

There are many in-vitro genotoxicity testing methods available. Some of the commonly used tests which are also a part of the standard battery are:

- Bacterial reverse mutation test which is otherwise called as Ames test whose endpoint is the gene mutations in the bacterial cell.

- Mammalian chromosome aberration test with the end point of chromosome aberration.

- Mammalian cell gene mutation test or the mouse lymphoma test whose end point is the gene mutations.

Bacterial reverse mutation test: The Bacterial reverse mutation test was developed by Ames. B thus the name Ames test. The amino acid requiring strains of Salmonella typhimurium and Escherichia coli are used in the bacterial reverse mutation test to detect the mutation points which may involve substitution, deletion or addition of one or few of base pairs of DNA. The main principle of the test is that after identifying the mutation it reverts it back and restore the functional capability of the mutant cell to synthesize Histidine. In this test the reverent bacteria cells are identified by the ability of the parent test strain to grow in the absence of amino acids. The bacterial reverse mutation test being rapid, inexpensive and easy to perform is commonly used as an initial screening test for genotoxicity or mutagenicity.

Mammalian chromosome aberration test: The main purpose of the mammalian chromosome aberration test is to identify the agents which can cause structural mutations in chromosomes or chromatids, chromatid mutation being the common. Other type of chromosomal changes like polyploidy and duplication can also be found using this test. A positive test result shows a potential mutagenic or carcinogenic of the agent but there is not a perfect correlation.

Mammalian cell gene mutation test: This test is used to find the gene mutations caused by the chemical substances. The commonly used cell lines include L5178Y mouse lymphoma cells, the CHO, CHO-AS52 and V79 lines of Chinese hamster cells, and TK6 human lymphoblastic cells. It detects the end points like thymidine kinase (TK) and hypoxanthine-guanine phosphoribosyltransferase (HPRT), and a transgene of xanthine-guanine phosphoribosyltransferase (XPRT) mutation.

In-vivo Genotoxicity Testing Methods

The in-vivo genotoxicity test or assays are done supplemental to in-vitro assay if an in-vitro positive result is obtained. Some of the in-vivo tests done are as follows:

In-vivo comet assay: It is one of the commonly used in-vivo test used for hazard assessment of agents which have potential for genotoxicity or mutagenicity . It helps in detecting the DNA damage and detects a broad variety of primary DNA lesions which cannot be identified by any other tests. This test can be applied to a wide variety of tissues or any special cell types. Being sensitivity to even low level of DNA damage it requires only small amount of cells per sample and it can be completed in a short period of time.

In-vivo micronuclei test/In-vivo chromosome aberration test: It is a test done to identify the damage done chromosome or spindles. On exposure to the mutagen the cell may undergo damage and on division it will form smaller micronucleus additional to the main nucleus.

Hepatotoxicity

Hepatotoxicity means damage to the liver caused by drugs and other factors resulting in problems in its functioning. Chemicals or drugs that cause hepatotoxicity are called hepatotoxins.

The liver is an organ that maintains the balance in our body. It is an organ where toxic compounds are detoxified, vitamins are stored, proteins, carbohydrates, and fats are broken down and bile acids are produced. The liver is also involved in molecular pathways involved in fighting infections, in reproduction, blood clotting and other functions. Any damage to the liver therefore can affect almost the entire body.

Drug-induced liver injury accounts for 5% of hospital admissions with jaundice and is the cause for nearly half of the acute liver failure cases. Women are normally more commonly affected by drug-induced liver injury compared with men.

The liver is the major organ in charge of breaking down fats, proteins, carbohydrates, drugs, and practically anything that enters our body. Since the metabolism of drugs occurs in the liver, the toxic effects of the drug affect the liver. Chemicals (e.g. drugs, smoking, alcohol) and toxins (plant, viral, or bacterial) induce changes to the biochemical pathways in the liver cells, or induce an immune response, or make the liver cells extremely sensitive to the defence missiles or cytokines in the body. The cells of the bile duct are are also affected by exposure to hepatotoxic drugs. Some patients progress to cirrhosis and liver failure, or can even develop benign or malignant cancers. The type of liver injury caused by chemicals can vary from:

- Hepatocellular injury, where the cells of the liver are damaged.

- Cholestatic injury which occurs due to obstruction to the flow of bile through the bile ducts. As a result, bile acids accumulate in the liver, which can damage it.

- Mixed, which has features of both hepatocellular and cholestatic injury.

- Other forms of liver injury like immunological injury, which occurs due to an immune reaction and mitochondrial injury, which can result in serious liver failure.

Factors that could increase the chance of hepatotoxicity by medications include:

- Presence of pre-existing liver disease like HIV, hepatitis B or hepatitis C, or due to alcohol intake.

- Presence of diseases affecting other organs like kidney failure.

- Intake of other liver-damaging medications or herbal products.

- Age, gender, genetic factors and ethnicity of the individual. Women appear to be affected more commonly.

Causes of Hepatotoxicity

Several drugs can be hepatotoxic depending on their dosage and the patient characteristics. The hepatotoxic drugs include:

- Non-steroidal anti-inflammatory drugs like acetaminophen and nimesulide

- Anti-retroviral drugs like zidovudine, didanosine and stavudine used in the treatment of HIV

- Antituberculosis drugs like rifampicin, isoniazid and pyrazinamide

- Antibiotics like erythromycin

- Cholesterol-lowering drugs like pravastatin, lovastatin and niacin

- Anesthetic agents like halothane

- Anti-epileptic drugs like carbamazepine, valproic acid, felbamate and phenytoin

- Anti-rheumatic drugs like methotrexate and azathioprine

- Cocaine

- Antipsychotics drugs like chlorpromazine and haloperidol

- Antidepressant drugs like amineptine and imipramine

- Anti-Alzheimer's drugs like tacrine

- Antihypertensive drugs like Methyldopa

- Oral contraceptives

- Hepatotoxicity may also be caused by environmental chemicals like arsenic, or intake of alcohol

Symptoms and Signs of Hepatotoxicity

There are 2 main types of hepatotoxicity – predictable (intrinsic) and idiosyncratic. Predictable injuries to the liver are dependent on the dose of the drug; they cause hepatotoxicity mainly at a higher dose. Idiosyncratic reactions are rare (1 in 1000 or 1 in 100 000 patients) and unpredictable drug reactions that occur at normal doses and often have serious consequences.

Symptoms of drug hepatotoxicity usually subside once the drug is withdrawn. They also depend on the type of liver injury and the degree of damage to the liver.

Symptoms of hepatotoxicity may include the following:

- Extreme exhaustion and fatigue

- Fever

- Jaundice: This is a classic symptom which causes the whites of the eyes, the skin, and the mucous membranes to become yellow.

- Severe itching

- Unusual increase in weight of over 1.5 to 2.5 kg in 1 week

- Swelling of feet and legs

- Nausea, vomiting, and pain in the abdomen

- Incessant bleeding in serious cases

Diagnosis of Hepatotoxicity

Hepatotoxicity may be diagnosed in the following ways:

Biopsy of the liver: A small tissue of the liver is removed and observed under the microscope to determine the type and extent of damage to the cells of the liver.

Endoscopic retrograde cholangiopancreatography (ERCP): This procedure is utilized to observe if the bile duct is blocked and is contributing to liver damage. As the name implies, an endoscope (a flexible tube with a light) is used to inject a dye into the biliary ducts. Any blockage in the bile duct is observed with the help of an x-ray.

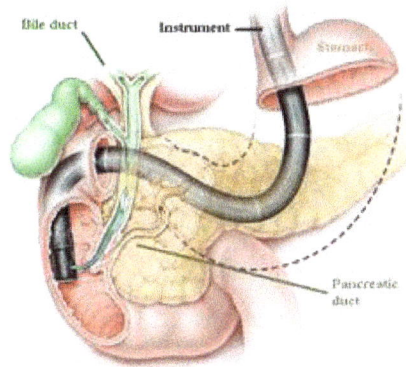

Blood tests:

- Serum ALT (alanine aminotransferase) activity: This is a very sensitive test and biomarker for hepatotoxicity. It is termed "the gold standard clinical chemistry marker of liver injury". The ALT enzyme is mainly found in the liver while minor quantities are observed in the skeletal muscle and heart. When the liver cells or hepatocytes are damaged, their contents are released into the bloodstream. Hence the levels of the ALT enzyme increase in the blood. Sometimes, injury to the skeletal muscle or heart can also drive up the levels of the ALT enzyme in the blood. Hence, there are chances of false positive results for hepatotoxicity in this test.

- Serum AST (aspartate aminotransferase) activity: The AST enzyme is present in the liver, heart, skeletal muscle, and brain. Like the ALT enzyme, the serum AST levels increase when hepatocytes are injured. Increased levels of the AST enzyme are also achieved when there is injury to organs other than the liver that contain the enzyme. This marker is not as specific as ALT. To confirm liver damage, the ratio of AST to ALT is measured.

There are other blood tests that are performed to supplement the results of the serum ALT test. They are listed below:

- Alkaline phosphatase activity: Any obstruction in the bile duct will lead to increased levels of alkaline phosphatase. Increased levels of ALP are a good indicator of drug-associated cholestasis in humans.

- Gamma-glutamyl transferase (GGT) levels: GGT levels in the blood also increase in cholestatic liver disease.

- Bilirubin and Bile acids: In the case of acute liver injury, bilirubin and bile acid levels may be increased in the blood.

- Other blood tests that are used to diagnose liver disease include measurement of serum albumin levels, serum 5¢ nucleotidase levels and prothrombin values.

Computed tomography scans (CT): In this test, the liver is scanned following injection of a radioactive dye into a vein. The three dimensional images obtained help to detect changes in the liver.

The history is an important diagnostic aspect of the diagnosis. The hepatotoxicity can more or less be assumed to be due to a drug if it occurred after intake of the medication, and if it improved after stopping the medication. It can be confirmed if the symptoms re-appear after re-instating the drug, but this is not done in clinical practice due to ethical reasons.

Treatment of Hepatotoxicity

The best way to avoid irreparable damage to the liver by a drug is to detect the toxic effects of the drug at its earliest stage, monitor and assess the damage, and decide on the line of treatment. Prompt attention will prevent the liver from being completely damaged to the point of no return.

The first step in treating hepatotoxicity caused by drugs is to stop taking the drug in question after consulting your doctor. Your doctor may prescribe medications to control certain symptoms. For example, the accumulation of fluid within the body can be reduced with diuretics. Cholestyramine may be used to treat itching.

N-acetylcysteine is prescribed for acetaminophen poisoning.

Liver transplantation may be required for severe liver injury.

Immunotoxicity

Immunotoxicity refers to the adverse effects on the structure or functioning of the immune system, or on other systems as a result of immune system dysfunction, caused by exposure to any toxic substance.

Adverse effects can be manifest as immunosuppression, hypersensitivity, or autoimmunity.

In any of these reactions, the cells, tissues, and organs of the immune system can be activated, inhibited, destroyed, or their responses exaggerated. It is the demonstrated involvement of at least one or more of these components that defines whether a consequence results from an actual immune response. For example, the term allergy should be used only to denote the involvement of immune responses, but is often mistakenly used to describe other adverse reactions.

An array of biological, physical, and chemical substances can perturb the intricate balance of the immune system, and this complex system provides several target sites for immunotoxicants. OTA broadly defines immunotixicant as a substance that leads to undesired effects on the immune system.

Whether a substance causes an adverse effect on the immune system, which could be permanent or reversible, depends on many factors, including the nature of the substance, dose and exposure, route and extent of exposure, the presence of other agents, and an individual's sex, genetic predisposition, age, and state of health.

While exposure to immunotoxic substances can occur anywhere to anyone, the opportunity to be exposed is greater for some populations of individuals than for others. Since immunotoxic effects have been described for some ambient and indoor air pollutants, some point out that the magnitude of individuals at risk could be tremendous.

Certain discrete populations are probably at increased likelihood of exposure, more amenable for scientific research, or both. For example, observations of occupational exposures to toxic sub stances have contributed much of the data pertinent to immunotoxicological effects — i.e., hypersensitivity, autoimmunity, and immunosuppression. Similarly, immunosuppressive drugs can be viewed as prototype immunotoxicants, and so patients receiving such drugs to prevent transplant rejection can be observed. Consumers of cosmetics have also been examined to detect hypersensitivity to various agents. Yet even among populations of individuals, not all of those exposed are affected. Certain individuals may be more susceptible to developing adverse reactions for a variety of reasons separate from mere exposure to the toxicant — even when dose and duration of exposure are similar for different individuals. For example, the person could be innately susceptible, i.e., be genetically predisposed. Age is another important factor in immunotoxicology, since the immune system is not completely functional at birth and might be more vulnerable to damage then. It also changes appreciably in older populations–generally to a less responsive level.

Finally, the overall state of health is also important in determining immunotoxicity.

Genetic background, in particular, is an important consideration in evaluating both potential immunotoxicity and variation in toxic effects. For example, genetic bases could account for the differential ability of individuals to produce anti-inflammatory factors, which in turn would influence how a person would be able to mount and modulate an immune response to an offending agent. Likewise, although almost anyone can develop an allergy to a given substance, a distinct segment (15 to 20 percent) of the population is clinically atopic, i.e., individuals who are unusually reactive to a variety of substances. Several genes have been identified that could influence this hyperreactivity, although specifics remain to be elucidated.

Not only is there a genetic basis for controlling the immune response to a particular antigen (as determined by histocompatibility or human leukocyte antigens (HLA), which present processed antigens, but other genes are responsible for certain pharmacologic abnormalities that can predispose an individual to immune dysfunctions, e.g., hyperreactivity to histamine or acetylmethylcholine. The latter substance is used to detect people with hyperreactive airways that constrict and cause wheezing when exposed to antigen. Total IgE production–central to immediate-type allergic reactions– is regulated by genes (. Persons could have genetic differences —singular or in combination— that might predispose them to allergies to certain environmental and occupational antigens. Certain HLA genes also are associated with increased risk of autoimmune diseases.

Finally, in addition to intrinsic factors, external risk factors are critical. Smoking and exposure to tobacco smoke, for example, are important external risk factors that must be accounted for in assessing potential immunotoxicity in humans. Another potential external risk factor is the presence of coincidental pulmonary infection at the time of exposure. The association of viral (but not bacterial) infection with triggering attacks of both childhood and adult onset asthma has been documented. Such infections can cause bronchial hyperreactivity that may even become permanent. It has been hypothesized that such an infection occurring in a worker simultaneous with exposure to an industrial substance or environmental pollutant could result in an enhanced sensitivity to that substance. A different external factor, house dust mites, is a precondition for certain childhood asthmas.

Immunotoxicity Testing

Two types of immunotoxicity tests/procedures are defined:

- Type 1 Tests are those that do not require any perturbation of the test animal, such as immunization and challenge with an infectious agent.

  i) Primary indicators of immune toxicity are derived from Basic Type 1 Tests, such as hematology and serum chemistry profiles, routine histopathology examinations, and organ and body weight measurements from standard toxicity. Additional procedures, such as measurements of thymus weights and performance of more definitive histopathological evaluations of immune-associated organs and tissues, have been added.

ii) Indicators of immune toxicity can also come from Expanded Type 1 Tests. These tests are logical extensions of Basic Type 1 tests; for example, Expanded Type 1 tests may extend the hematology, serum chemistry, and histopathology evaluations of standard toxicity studies. Many of these expanded tests can be performed with the same blood and tissue samples collected for the Basic Type 1 tests; in addition, many of the expanded tests can be performed retrospectively.

- Type 2 Tests include injections or exposure to test antigens, vaccines, infectious agents or tumor cells. If Type 2 tests are to be performed concurrently with a standard toxicity study, a satellite group of animals should be added to the recommended number of test animals in the study. Protocol designs for standard toxicity studies that include a satellite group of animals for Type 2 immunotoxicity tests will be recommended when available information indicates that a test compound may present an immunotoxic risk.

Sets of Basic and Expanded Type 1 Tests are defined as Level I Immunotoxicity Tests. Some Level I tests screen for immunotoxic effects in test animals; others focus on de fining an immunotoxic effect more specifically, such as determining the mechanism or cell types involved. Analogously, sets of Type 2 tests are defined as Level II Immunotoxicity Tests; Level II tests also can be used to screen for, or more specifically define, immunotoxic effects of food and color additives used in food.

Indicators of Possible Immune Toxicity

Basic Type 1 Tests: Primary Indicators

The primary indicators of possible immune toxicity are derived from routine measurements and examinations performed in toxicity. Indicators derived from short-term and subchronic toxicity studies, and developmental toxic ity studies with rodents are listed below. If a substance produces one or more of these primary indicators of immune toxicity, more definitive immunotoxicity tests (Expanded Type 1 tests or Type 2 tests) may be recommended; such decisions will be made on a case-by-case basis.

Indicators from Short-term and Subchronic Toxicity Studies

Hematology Indicators: Elevation or depression in white blood cell (WBC) counts; altered differential WBC counts; lymphocytosis and lymphopenia; and eosinophilia.

Clinical Chemistry Indicators: Elevated or reduced total serum protein in combination with an abnormal albumin-to-globulin (A/G) ratio. Other indicators often associated with immunologic dysfunction include abnormal levels of liver proteins and enzymes, such as albumin and the transaminases.

Histopathology Indicators: Abnormalities found during gross and routine histological evaluation of the lymphoid tissues, e.g. spleen, lymph nodes, thymus, gut-associated lymphoid tissue (GALT, in particular Peyer's patches), and bone marrow. Morphologic abnormalities such as scattered, focal mononuclear cell infiltrates in non-lymphoid organs (e.g. kidney and liver) may be relevant to autoimmune disease. If differences are seen in any lymphoid tissue, attention should be given to "cellularity" and prevalence of activated macrophages. The description could include in situ descriptions of the types of cells, density of the cell populations, lymphocyte distribution relative to distinguishing structures or defined areas of the organ. (In these instances, the effect does not need to be defined rigorously for each animal. The number of animals observed, however, should be a statistically significant sample size.) The histopathological analysis of routinely stained (hematoxylin and eosin) samples of the spleen should include descriptions of lymphocyte distribution and proliferation in known T- and B-cell areas, such as the germinal centers (for B-cells) and the periarteriolar lymphocyte sheath (PALS) for Tcells if abnormalities are observed. The histopathologic analysis of the lymph nodes and Peyer's patches should include a description of the immune activation (i.e. the relative number of follicles and germinal centers) when abnormalities or lesions are observed in these organs. When abnormalities of the thymus are observed, histopathologic analysis should be descriptive and quantitative as possible with regard to atrophy and necrosis and other observations. If the test compound is shown to either stimulate cell proliferation, or to cause atrophy and cell depletion in any lymphoid organ, the effect is likely to be viewed as a potentially immunotoxic effect requiring more definitive testing.

Organ and Body W eight Indicators: Elevated or depressed spleen and thymus weights; elevated or depressed organ-to-body-weight ratios for the spleen and thymus (statistical treatment of the organ-to body-weight ratios should include an analysis of co-variance, with body weight as the co-variant).

Elevated or depressed body weights, although primarily an indicator of endocrine function, may also indicate indirect immunotoxic effects, since endocrine function can significantly effect the immune system.

Indicators from Developmental Toxicity Studies

Morbidity and Mortality Indicators: Unusual incidence of maternal infections.

Histopathology Indicators: Abnormalities found during gross evaluation of the fetal liver, spleen, and thymus.

For animals in the F1 and F2 generations:

Hematology Indicators: Elevation or depression in white blood cell (WBC) counts; altered differential WBC counts; lymphopenia and lymphocytosis; and eosinophilia.

Clinical Chemistry Indicators: Eleva ted or reduced total serum protein in combination with an abnormal albumin-to-globulin (A/G) ratio.

Histopathology Indicators: Abnormalities found during gross and routine histological evaluation of the lymphoid tissues, e.g. spleen, lymph nodes, thymus, gut-associated lymphoid tissue (GALT, in particular Peyer's patches), and bone marrow. Morphologic abnormalities such as scattered, focal mono-nuclear cell infiltrates in non-lymphoid organs (e.g. kidney and liver) may be relevant to autoimmune disease. If differences are seen in any lymphoid tissue, attention should be given to "cellularity" and prevalence of activated macrophages. The description could include in situ descriptions of the types of cells, density of the cell populations, lymphocyte distribution relative to distinguishing structures or defined areas of the organ. (In these instances, the effect does not need to be defined rigorously for each animal. The number of animals observed, however, should be a statistically significant sample size.) The histopathological analysis of routinely stained (hematoxylin and eosin) samples of the spleen should include descriptions of lymphocyte distribution and proliferation in known T- and B-cell areas, such as the germinal centers (for B-cells) and the periarteriolar lymphocyte sheath (PALS) for T-cells if abnormalities are observed. The histopathologic analysis of the lymph nodes and Peyer's patches should include a description of the immune activation (i.e. the relative number of follicles and germinal centers) when abnormalities or lesions are observed in these organs. When abnormalities of the thymus are observed, histopathologic analysis should be descriptive and quantitative as possible with regard to atrophy and necrosis and other observations. If the test compound is shown to either stimulate cell proliferation, or to cause atrophy and cell depletion in any lymphoid organ, the effect is likely to be viewed as a potentially immunotoxic effect requiring more definitive testing.

Organ and Body W eight Indicators: Elevated or depressed spleen and thymus weights; elevated or depressed organ-to-body-weight ratios for the spleen and thymus (statistical treatment of the organ-to-body-weight ratios should include an analysis of co-variance, with body weight as the co-variant). Elevated or depressed body weights, although primarily an indicator of endocrine function, may also indicate indirect immunotoxic effects, since endocrine function can significantly effect the immune system.

Expanded Type 1 Immunotoxicity Tests

Assessing the safety of food and color additives used in food usually requires the completion of various toxicity studies. In addition to the screen of primary indicators of possible immune toxicity provided by these toxicity studies and summarized above, additional tests for further evaluation of the immunotoxic potential of a test substance may be recommended by the Agency. The Agency's recommendation that specific immunotoxicity tests be performed on test substances will be made on a case-by-case basis. Expanded Type 1 immunotoxicity tests include:

- Hematology Tests: Flow cytometric analysis of B-lymphocytes, T-lymphocytes, and T-lymphocyte subsets (TH + TS or CD4 and CD8); immunostaining (immunoperoxidase or immunofluorescence) of Blymphocytes, T-lymphocytes and T-lymphocyte subsets from peripheral blood or single cell suspensions from the spleen.

  Hematology Indicators: Decreased or elevated percentages of any of the various lymphocytes relative to controls and abnormalities in the B-cell/T-cell and the TH/TS (CD4/CD8) cell ratios; these should be determined from differential counts of the immunostained preparations or from cytometric analysis.

- Serum Chemistry Tests: Electrophoretic analysis of serum proteins to permit separation and quantification of the relative percentages of albumin and the "-, $-, and J-globulin fractions; quantification of J-globulin fractions (IgG, IgM, IgA, and IgE); analysis of total serum complement and components of complement (such as C3) from CH-50 determinations; immunochemical assay of serum cytokines, such as IL-2, IL-1, and J-interferon; quantification of serum auto-antibodies, such as antinuclear, anti-mitochondrial, and anti-parietal cell antibodies.

  Serum Chemistry Indicators: Statistically significant variations between experimental and control groups of animals for any of the parameters listed above.

- Histopathology Tests: Immunostaining of B-lymphocytes in the spleen and lymph nodes, using polyclonal antibodies to IgG of the test anima ls; 30,31 immunostaining of T-lymphocytes and Tlymphocyte subsets in the spleen, using monoclonal or polyclonal antibodies to various cell markers; micro-metric measurements of germinal centers and PALS of the spleen and the follicles and germinal centers of lymph nodes; morphome tric analysis of the relative areas of the cortex and medulla of the thymus, using routinely stained histopathology sections.

  Histopathology Indicators: Statistically significant variations between experimental and control groups of animals for any of the parameters listed above, using both analysis of variance (ANOVA) and a multiple comparison T-test, such as Dunnett's.

- Tests for In Vitro Analysis of the Functional Capacity of Specific Cell Types:

  Activity of Natural Killer (NK) Cells: The functional capacity of NK cells can be measured using the classical 51Cr chromium release assay; 19 this assay is well standardized and has been used successfully with both mice and rats in various immunotoxicity studies. 33-35 Of particular concern is reduced NK cell activity, which may be correlated with increased tumorigenesis and infectivity.

Mitogenic Stimulation Assays for B- and T-Lymphocytes: Certain plant lectins stimulate blastogenesis and DNA synthesis of T- and B-lymphocytes: concanavalin-A (Con-A) and phytohemagglutinin (PHA) are known to preferentially stimulate T-lymphocytes, and an extract from pokeweed (PWM) as we ll as certain bacterial lipopolysaccharides (LPS) and protein extracts are known to preferentially stimulate B-lymphocytes in vitro. Since these assays are carried out ex vivo, they can be performed on preparations of peripheral blood. The assays are well characterized for use in various animals species (including man36), can be performed on either peripheral blood or spleen-cell suspensions, and have been used in a number of immunotoxicity studies. Both reduced and e levated levels of blastogenesis or 3H incorporation into DNA are of interest in the evaluation of the immunotoxic potential of food and color additives used in food.

Phagocytotic Index of the Macrophage: Various assays to determine the phagocytotic ability of macrophages have been described. These assays measure the ability of a macrophage to ingest particulate substances, such as plastic beads or iron filings, and can be performed on peripheral blood or single cell suspensions of lymphoid organs, such as the spleen. Other assays measure the capacity of the macrophage to destroy live bacteria through lysosomal enzyme activity.

Stem Cell Assays: Bone marrow preparations can be used to investigate the pluripotent population or specific progenitor populations. 41 Although these assays have not been used extensively in immunotoxicity evaluations, they may be recommended when histopathological evaluation indicates that the test substance may have caused changes in bone marrow.

Type 2 Immunotoxicity Tests

Evaluating the functional capacity of the immune system requires injecting a substance that elicits immunological reactivity in a test animal. Various antigens provide information about the types of immunity or cells that may be involved in an immune response. For example, protein antigens usually elicit T-dependent immune responses with subsequent production of antibodies to the protein. Polysaccharides elicit T-independent immune responses. Some antigens elicit cell-mediated immune responses, while immunogens such as complex bacteria and viruses may elicit humoral and cell-mediated responses. All of the antigens listed below have been tested in rodents; when an antigen has been used preferentially with a particular rodent species, this is noted.

- T-Dependent Test Antigens: One of the most widely used antigens for rodents and nonrodents is sheep red blood c ells (SRBC). For example, SRBCs have been widely used in mice in the Plaque-Forming Cell Assay: antibody-produc ing spleen cell suspensions are mixed with SRBCs, placed on covered slides, and incubated; each antibody-producing cell causes a small, clear area (plaque) to form on the slide; the plaques are then counted. Other T-dependent test antigens that have been widely used include keyhole limpet hemocyanin and bovine serum albumin.

- T-Independent Test Antigens: Ficoll, a branched chain polysaccharide, haptenated ficoll, polyvinylpyrrolidone, and bacterial lipopolysaccharides have been used as T-independent test antigens with mice and rats.

- Human Vaccines: Human T-dependent vaccines, such as tetanus toxoid, and the T-independent vaccine containing pneumococcal polysaccharide antigens have been used in both rats and mice. It is possible to compare responses of the test speciies to the vaccines with human responses, because standard human sera are available from FDA's Center for Biologics.

- Test Antigens for Cell-Mediated Immune (CMI) Reactivity: Contact sensitizers such as dinitrochlorobenzene (DNCB) have been used to elicit delayed hypersensitivity (DTH) responses as a measure of CMI in animals. These assays can be performed in rodent as well as non-rodent species. The DTH assays are economical and correlate well with decreased CMI and host resistance to infectious agents in humans, as well as animals. The mixed-lymphocyte response (MLR) assay, which uses lymphocytes from a different strain, has been successfully used to evaluate CMI in mice.

- Host Resistance Assays with Infectious Agents: A number of bacterial strains have been used to measure host resistance, including Listeria monocytogenes, various strains of Streptococcus, and Escherichia coli. Useful viral models include influenza, herpes, and cytomegalovirus. A yeast infectivity model using Candida albicans has been described, as well as parasitic infectivity models using Trichinella spiralis and Plasmodium yoelli.

- Host Resistance Assays Using Syngeneic Tumor Cells: Various assays of host resistance have been described using a number of cultured tumor cell lines. These assays, unlike those involving the infectious agents discussed above, do not require special barrier facilities to prevent infections from spreading throughout an animal colony. Two mouse assays have been validated: the PYB6 sarcoma assay and the B16F10 melanoma assay. An assay using a lung tumor model and the MADB106 tumor cell line also has been validated for use in immunotoxicity studies.

Nephrotoxicity

Nephrotoxicity is one of the most common kidney problems and occurs when your body is exposed to a drug or toxin that causes damage to your kidneys. When kidney damage occurs, you are unable to rid your body of excess urine, and wastes. Your blood electrolytes (such as potassium, and magnesium) will all become elevated.

Nephrotoxicity can be temporary with a temporary elevation of lab values (BUN and/ or creatinine). If these levels are elevated, these may be due to a temporary condition such as dehydration or you may be developing renal (kidney failure). If the cause of the increased BUN and creatinine levels is determined early, and your healthcare provider implements the appropriate intervention, permanent kidney problems may be avoided.

Nephrotoxicity may also be referred to as renal toxicity.

Nephrotoxicity or renal toxicity can be a result of hemodynamic changes, direct injury to cells and tissue, inflammatory tissue injury, and obstruction of renal excretion. Nephrotoxicity is frequently induced by a wide spectrum of therapeutic drugs and environmental pollutants. Knowledge of the complex molecular and pathophysiologic mechanisms leading to nephrotoxicity remains limited, in part, by research that historically focused on single or relatively few risk markers. As such, current kidney injury biomarkers are inadequate in terms of sensitivity and specificity. In contrast, metabolomics enables screening of a vast array of metabolites simultaneously using NMR and MS to assess their role in nephrotoxicity development and progression. A more comprehensive understanding of these biochemical pathways would also provide valuable insight to disease mechanisms critical for drug development and treatment.

Nephrotoxicity is one of the key mechanisms by which pharmacodynamic drug interactions occur with AmB. Three types of interactions may result from AmB nephrotoxicity:

- Synergistic nephrotoxicity such as with cyclosporin and tacrolimus, cisplatin, and antimicrobials (aminoglycosides, high-dose trimethoprim-sulfamethoxazole, foscarnet and cidofovir). It is noteworthy that this synergistic toxicity may be precipitated by even a single dose of these agents, particularly when they are added in patients receiving calcineurin inhibitors.

- Delayed clearance of other renally excreted drugs as a result of AmB nephrotoxicity, hence causing accumulation of these agents with potential toxicity (melphalan, other antineoplastic agents, ganciclovir, 5FC).

- Electrolyte abnormalities that may either increase the toxicity of concomitant agents (hypokalemia in patients receiving digoxin) or may be worsened by co-administration of agents known to cause electrolyte abnormalities (foscarnet).

Neurotoxicity

The term neurotoxicity refers to damage to the brain or peripheral nervous system caused by exposure to natural or man-made toxic substances.

These toxins can alter the activity of the nervous system in ways that can disrupt or kill nerves. Nerves are essential for transmitting and processing information in the brain, as well as other areas of the nervous system.

Due to their high metabolic rate, neurons are at the greatest risk of damage caused by neurotoxins. This is followed, in order of risk, by oligodendrocytes, astroocytes, microglia and capillary endothelium cells.

Depending on a neurotoxin's chemical profile, it will cause damage to certain parts or particular cellular elements of the nervous system. Non-polar substances are more soluble in lipids and can therefore access the nervous tissue more easily than polar compounds, which are less soluble in lipids. The body's response to neurotoxins is influenced by factors such as the neurotransmitter affected, cellular membrane integrity and the presence of detoxifying mechanisms.

Some examples of substances that can be neurotoxic to humans include:

- Chemotherapy drugs that are used to kill fast growing cells
- Radiation
- Drug therapies or drugs of abuse
- Heavy metals such as mercury and lead
- Certain foods and food additives
- Insecticides/pesticides
- Cosmetics
- Industrial and cleaning solvents

Some examples of neurotoxic substances our environment has become polluted with and that it is difficult for people to avoid exposure to include:

- Mercury
- Cadmium
- Lead
- Insecticides
- Solvents
- Car exhaust
- Chlorine
- Formaldehyde
- Phenol

Effects of Neurotoxicity

Some of the effects of neurotoxicity may appear immediately, while others can take months or years to manifest.

The effects of neurotoxicity depends on various different factors such as the characteristics of the neurotoxin, the dose a person has been exposed to, ability to metabolize and excrete the toxin, the ability of affected mechanism and structures to recover and how vulnerable a cellular target is.

Some of the symptoms of neurotoxicity include:

- Paralysis or weakness in the limbs
- Altered sensation, tingling and numbness in the limbs
- Headache
- Vision loss
- Loss of memory and cognitive function
- Uncontrollable obsessive and compulsive behavior
- Behavioral problems
- Sexual dysfunction
- Depression
- Loss of circulation
- Imbalance
- Flu-like symptoms

Other conditions that may develop as a result of neurotoxicity include chronic fatigue syndrome, attention deficit hyperactivity disorder, chronic sinusitis and asthma that does not respond to therapies. Symptoms may also resemble those seen in some auto-immune conditions such as irritable bowel syndrome or rheumatoid arthritis.

Some examples of toxins that occur naturally in the brain and can lead to neurotoxicity include oxygen radicals, beta amyloid and glutamate. Aside from causing movement disorders, cognitive deterioration and dysfunction of the autonomic nervous system, neurotoxicity has also been shown to be a major contributor to progressive neurological disorders such as Alzheimer's disease.

The fundamental purpose of testing chemical substances for neurotoxicity is to prevent disease by identifying toxic hazards before humans are exposed. That approach to disease prevention is termed "primary prevention." In contrast, "secondary prevention" consists of the early detection of disease or dysfunction in exposed persons and populations followed by prevention of additional exposure.

In the most effective approach to primary prevention of neurotoxic disease of environmental origin, a potential hazard is identified through premarket testing of new chemicals before they are released into commerce and the environment. Identifying potential neurotoxicity caused by the use of illicit substances of abuse or by the consumption of foods that contain naturally occurring toxins is less likely. Disease is prevented by restricting or banning the use of chemicals found to be neurotoxic or by instituting engineering controls and imposing the use of protective devices at points of environmental release.

Each year, 1,200–1,500 new substances are considered for premarket review by the Environmental Protection Agency (EPA) and several hundred compounds are added to the 70,000 distinct chemicals and the more than 4 million mixtures, formulations, and blends already in commerce. The proportion of the new chemicals that could be neurotoxic if exposure were sufficient is not known and cannot be estimated on the basis of existing information. However, of the 588 chemicals used in substantial quantities by American industry in 1982 and judged to be of toxicologic importance by the American Conference of Governmental Industrial Hygienists (ACGIH), 28% were recognized to have adverse effects on the nervous system; information on the effects was part of the basis of the exposure limits recommended by ACGIH.

Given the absence of data on neurotoxicity of most chemicals, particularly industrial chemicals, it is clear that comprehensive primary prevention would require an extensive program of toxicologic evaluation. EPA has regulatory mechanisms to screen chemicals coming into commerce, but overexposures continue to occur, and episodes of neurotoxic illness have been induced by chemicals, such as Lucel-7, that have slipped through the regulatory net. Serious side effects of pharmaceutical agents also continue to surface, such as the 3 million cases of tardive dyskinesia that developed in patients on chronic regimens of antipsychotic drugs. It is now possible to identify only a small fraction of neurotoxicants solely on the basis of chemical structure through analysis of structure-activity relationships (SARs), so in vivo and in vitro tests will be needed for premarket evaluation until greater understanding of SARs permits them to be used with confidence.

If neurotoxic disease is to be prevented, public policy must be formulated as though all chemicals are potential neurotoxicants; a chemical cannot be regarded as free of neurotoxicity merely because data on its toxicity are lacking. Prudence dictates that all chemical substances, both old and new, be subjected to at least basic screening for neurotoxicity in the light of expected use and exposure. However, the sheer number of untested chemicals constitutes a practical problem of daunting magnitude for neurotoxicology. Given the number of untested chemicals and current limitations on resources, they cannot all be tested for neurotoxicity in the near future. Testing procedures designed for neurotoxicologic evaluation that have been developed so far might be reasonably effective, but are so resource-intensive that they could not be applied to all untested chemicals.

A rational approach to neurotoxicity testing must contain the following elements:

- Sensitive, replicable, and cost-effective neurotoxicity tests with explicit guidelines for evaluating and interpreting their results.

- A logical and efficient combination of tests for screening and confirmation.

- Procedures for validating a neurotoxicologic screen and for guiding appropriate confirmatory tests.

- A system for setting priorities for testing.

Approach to Neurotoxicity Testing

Difficulties in Neurotoxicity Testing

Neurotoxicity testing is relatively new. Although its rapid development is noteworthy, its progress has been constrained by several factors that complicate neurotoxicologic assessment. Some of the complexities, such as sex- or age-related variability in response, are common to all branches of toxicology. Neurotoxicology, however, faces unique difficulties, because of several characteristics that make the nervous system particularly vulnerable to chemically induced damage. Those characteristics include the limited ability of the nervous system to repair damage, because of the absence of neurogenesis in adults; the precarious dependence of axons and synaptic boutons on long-distance intracellular transport; the system's distinct metabolic requirements; the system's highly specialized cellular subsystems; the use of large numbers of chemical messengers for interneuronal communication; and the complexity of the nervous system's structural and functional integration. The nervous system exhibits a greater degree of cellular, structural, and chemical heterogeneity than other organ systems. Toxic chemicals potentially can affect any functional or structural component of the nervous system—they can affect sensory and motor functions, disrupt memory processes, and cause behavioral and neurologic abnormalities. The large number of unique functional subsystems suggests that a great diversity in test methods is needed to ensure assessment of the broad range of functions susceptible to toxic impairment. The special vulnerability of the nervous system during its long period of development is also a critical issue for neurotoxicology.

Despite the inherent difficulties of neurotoxicity testing, some validated tests have been developed and implemented. Testing strategies must take those facets of the nervous system into account, and they must consider a number of variables known to modify responses to neurotoxic agents, such as the developmental stage at which exposure occurs and the age at which the response is evaluated. The issue of timing is complex. During brain development, limited damage to cell function—even reversible inhibition of transmitter synthesis, for instance—can have serious, long-lasting effects, because of the trophic functions of neurotransmitters during neuronal development and synaptogenesis. Rodier Kellogg and others have shown the striking differences between the effects of perinatal exposure of animals to some neuroactive drugs (e.g., diazepam) and the effects of exposure of adult animals to the same drugs. Other agents (e.g., methylmercury and lead) are toxic at every age, but are toxic at lower doses in developing organisms. In addition, in developing animals, the blood-brain barrier might not be sufficiently developed to exclude toxicants. Stresses later in development might lead to the expression of relatively sensitive effects that were latent or unchallenged at earlier stages of development. During senescence, the CNS undergoes further change, including a loss of nerve cells in some regions. CNS function in senescence could be vulnerable to cytotoxic agents that, if encountered earlier in development, might have been protected against by redundant networks or compensated for by "rewiring" of networks. To address those complex issues, testing paradigms that incorporate both exposures and observations during development and during aging need to be considered. The disorders might be acute and reversible or might lead to progressive disorders over the course of chronic exposure. More sensitive biologic markers as early indicators of neurotoxicity are urgently needed. Opportunities should be exploited for detecting neurotoxicity when chronic lifetime bioassays are conducted for general toxicity or carcinogenicity.

In the design of neurotoxicity screens, no test can be used to examine all aspects of the nervous system. The occurrence of an effect of a chemical on one function of the nervous system will not necessarily predict an effect on another function. Therefore, it is important to use a variety of initial tests that measure different chemical, structural, and functional changes to maximize the probability of detecting neurotoxicity or to use tests that sample many functions in an integrated fashion.

The Testing Strategy

Efficient identification of potential hazards warrants a tiered testing strategy. The first tier of testing (the screen) need not necessarily be specifically predictive of the neurotoxicity likely to occur in humans, unless regulatory agencies are to use the results for direct risk-management decisions. The tests in later tiers are essential to assess specificity and confirm screening results and are appropriate for defining dose-response relationships and mechanism of action. Screening tests would be followed, as appropriate, by more specific assessments of particular functions. Such an approach permits a decision to be made about whether to continue testing at each step of the progression. In the case of chemicals already in use (in which case people are already being exposed and financial consequences might be considerable), detailed testing to determine mechanisms of toxicity would be pursued when screening tests revealed neurotoxic effects; positive findings on a chemical undergoing commercial development might trigger its abandonment with no further testing.

Testing at the first tier is intended to determine whether a chemical has the potential to produce any neurotoxic effects—i.e., to permit hazard identification, the first step of the NRC (1983) risk-assessment paradigm. The next tier is concerned with characterization of neurotoxicity, such as the type of structural or functional damage produced and the degree and location of neuronal loss. During hazard characterization (the second step of the NRC paradigm), tests are used to study quantitative relationships between exposure (applied dose) and the dose at the target site of toxic action (delivered dose) and between dose and biologic response. The third and final tier of neurotoxicity testing is the study of mechanism(s) of action of chemical agents.

The decision to characterize a chemical through second-tier testing might be motivated by structure-activity relationships, existing data that suggest a chemical is neurotoxic, or reports of neurotoxic effects in humans exposed to the chemical, in addition to the results of first-tier testing. Testing at this second tier can help to resolve several issues, including whether the nervous system is the primary target for the chemical and what the dose-effect and time-effect relationships are for relatively sensitive end points. Such tests can also be useful in the determination of a no-observed-adverse-effect level (NOAEL) or lowest-observed-adverse-effect level (LOAEL). Experiments done at the third tier can also examine the mechanism of action associated with a neurotoxic agent; they will often involve neurobehavioral, neurochemical, neurophysiologic, or neuropathologic measures. They might also suggest biologic markers of neurotoxicity for validation and use in new toxicity tests and in epidemiologic studies.

Characteristics of Tests useful for Screening

Like any other toxicity test, screening tests for neurotoxicity should be sensitive, specific, and valid. Sensitivity is a test's ability to detect an effect when it is produced (the ability to register early or subtle effects is especially desirable). Sensitivity depends on inherent properties of the test and on study design factors, such as the numbers of animals studied and the amount and duration of exposure. Specificity is a test's ability to respond positively only when the toxic end point of interest is present. Specificity and sensitivity are aspects of accuracy. An inaccurate test fails to identify the hazardous potential of some substances and incorrectly identifies as hazardous other substances that are not. In statistical terms, the failure to identify a hazard is a false-negative result, and the mistaken classification of a safe substance as hazardous is a false-positive result. Increasing test specificity reduces the incidence of false positives, but often has the unwanted consequence of increasing the incidence of false negatives (decreasing sensitivity); increasing test sensitivity reduces the incidence of false negatives, but often increases the incidence of false positives.

An ideal screen would have broad specificity, so that it could detect all aspects of nervous system dysfunction. In practice, no screening procedure is likely to provide the desired coverage without producing false positives. The sensitivity of the first tier might be maximized by a battery of screening tests that are individually quite specific. Several durations of exposure, long postexposure observation periods, and lifetime tests might all be necessary to cover the possible manifestations of neurotoxicity. In a tiered test system, very high sensitivity of the screening tier is usually considered essential; the loss of specificity must be compensated for in later tiers, to reveal the false positives. If, however, product development will be aborted on any indication of neurotoxicity, false positives could have high social costs. In the later tiers, narrower specificity is appropriate to characterize a suspected toxicant or to establish its mechanism of action.

Validation is the process by which the credibility of a test is established for a specific, purpose. It entails demonstrating the reliability of the test's performance in giving reproducible results within a laboratory and in different laboratories and giving appropriate results for a control panel of substances of known toxicity. The usefulness of the results of a neurotoxicity screening test depends on a positive outcome's being strongly correlated with a neurotoxic effect that is actually caused by exposure to the test substance. Direct mechanistic causality is not essential for their interpretation, although direct insight into mechanism would be valuable, as is sought when developing a biologic marker of effect.

Validation of a test system should also include demonstration that positive test results indicate that neurotoxic effects would occur if humans were exposed to the substance. Predicting an influence on human affect or cognition with nonhuman test systems is challenging, but possible. Many animal testing models do produce effects that correspond to those seen in humans exposed to the same substance, but a result would also be considered valid if it correctly predicted any neurotoxic effect in the human population. For example, consider a screening test for an effect produced only at high doses that is consistently correlated with a milder, presumably precursor effect that occurs at lower doses. The easily detectable effect occurring at high doses in experimental animals might never be observed in humans, whose exposure would never be extreme, but its occurrence in the animal model might indicate that low-level exposure in humans could produce a more subtle toxic effect. O'Donoghue has noted that chemicals that damage axons are often metabolic poisons that produce a retardation of weight gain without decreasing food intake—a more easily observed end point. It is important to recognize that chemicals can adversely affect multiple organ systems, and an effect in other organs might influence some measures of neurotoxicity. In vitro assay systems often measure end points that appear unrelated to neurologic functioning in whole animals, but detect alterations to crucial underlying mechanistic processes.

Application of the Screen and Later Tiers

Given the enormous number of substances that have not been tested for neurotoxicity (or any form of toxicity), some characteristics of chemicals, such as their structure or production volume, must contribute to the determination of priority for screening. Confidence in the combined sensitivity of tests composing a screening battery should be high enough for negative findings to be reliably regarded as acceptable evidence that a substance is unlikely to have neurotoxic activity without the need for additional testing. If a new chemical gives positive results, there might nevertheless be academic or commercial interest in pursuing a risk assessment, or its development for use might be abandoned without further testing. Detailed testing to characterize neurotoxicity revealed by screening procedures will be more common for existing substances with commercial value or wide exposure. In either case, the path followed through the tiers of testing would be contingent on a chemical's unfolding toxicity profile, including types of toxicity other than neurotoxicity. A feasible screening system must strike a balance among the amount of time and expense that society is willing to expend in testing, its desire for certainty that neurotoxic substances are being kept out of or removed from the environment, and its interest in gaining benefits from various types of chemicals.

Treatment

The treatment approach to neurotoxicity is elimination or reduction of the toxic substance and therapy to relieve symptoms or provide support. Treatment may also involve avoiding air, food and water pollutants. Some examples of therapies used in the treatment of neurotoxicity include massage, exercise and immune modulation.

Prognosis

The outcome of neurotoxicity depends on the duration and extent of exposure to the toxic substance, as well as the degree of neural damage. Exposure to neurotoxins can be fatal in some cases, while in others, patients survive but may not completely recover. In other cases, patients do completely recover after receiving treatment.

Reproductive Toxicity

Reproductive toxicity is defined as adverse effects of a chemical substance on sexual function and fertility in adult males and females, as well as developmental toxicity in the offspring.

In this classification system, reproductive toxicity is subdivided under two main headings:

Adverse Effects on Reproductive Ability or Capacity

Any effect of chemicals that would interfere with reproductive ability or capacity. This may include, but not be limited to, alterations to the female and male reproductive system, adverse effects on onset of puberty, gamete production and transport, reproductive cycle normality, sexual behavior, fertility, parturition, premature reproductive senescence, or modifications in other functions that are dependent on the integrity of the reproductive systems. Adverse effects on or via lactation can also be included in reproductive toxicity, but for classification purposes, such effects are treated separately. This is because it is desirable to be able to classify chemicals specifically for adverse effect on lactation so that a specific hazard warning about this effect can be provided for lactating mothers.

Adverse Effects on Development of the Offspring

Taken in its widest sense, developmental toxicity includes any effect which interferes with normal development of the conceptus, either before or after birth, and resulting from exposure of either parent prior to conception, or exposure of the developing offspring during prenatal development, or postnatally, to the time of sexual maturation. However, it is considered that classification under the heading of developmental toxicity is primarily intended to provide hazard warning for pregnant women and men and women of reproductive capacity. Therefore, for pragmatic purposes of classification, developmental toxicity essentially means adverse effects induced during pregnancy, or as a result of parental exposure. These effects can be manifested at any point in the life span of the organism. The major manifestations of developmental toxicity include:

(1) Death of the developing organism,

(2) Structural abnormality,

(3) Altered growth,

(4) Functional deficiency.

Considerations

The purpose of the harmonized system for the classification of chemicals which may cause an adverse effect on reproduction in humans is to provide a common ground which could be used internationally for the classification of reproductive toxicants.

The system is hazard based, classifying chemicals on the basis of intrinsic ability to produce an adverse effect on reproductive function or capacity, and on development of the offspring. The present system involves consideration of any substance-related adverse effect on reproduction seen in humans, or observed in appropriate tests conducted in experimental animals.

Classification Criteria for Substances

Weight of Evidence

Classification as a reproductive toxicant is made on the basis of an assessment of the total weight of evidence. This means that all available information that bears on the determination of reproductive toxicity is considered together. Included are such information as epidemiological studies and case reports in humans and specific reproduction studies along with sub-chronic, chronic and special study results in animals that provide relevant information regarding toxicity to reproductive and related endocrine organs. Evaluation of substances chemically related to the material under study may also be included, particularly when information on the material is scarce. The weight given to the available evidence will be influenced by factors such as the quality of the studies, consistency of results, nature and severity of effects, level of statistical significance for intergroup differences, number of endpoints affected, relevance of route of administration to humans and freedom from bias. Both positive and negative results are assembled together into a weight of evidence determination. However, a single, positive study performed according to good scientific principles and with statistically or biologically significant positive results may justify classification.

Toxicokinetic studies in animals and humans, site of action and mechanism or mode of action study results may provide relevant information, which could reduce or increase concerns about the hazard to human health. If it can be conclusively demonstrated that the clearly identified mechanism or mode of action has no relevance for humans or when the toxicokinetic differences are so marked that it is certain that the hazardous property will not be expressed in humans then a substance which produces an adverse effect on reproduction in experimental animals should not be classified.

In some reproductive toxicity studies in experimental animals the only effects recorded may be considered of low or minimal toxicological significance and classification may not necessarily be the outcome. These include for example small changes in semen parameters or in the incidence of spontaneous defects in the fetus, small changes in the proportions of common fetal variants such as are observed in skeletal examinations, or in fetal weights, or small differences in postnatal developmental assessments.

Data from animal studies ideally should provide clear evidence of specific reproductive toxicity in the absence of other, systemic, toxic effects. However, if developmental toxicity occurs together with other toxic effects in the dam, the potential influence of the generalized adverse effects should be assessed to the extent possible. The preferred approach is to consider adverse effects in the embryo/fetus first, and then evaluate maternal toxicity, along with any other factors which are likely to have influenced these effects, as part of the weight of evidence. In general, developmental effects that are observed at maternal toxic doses should not be automatically discounted. Discounting developmental effects that are observed at maternal toxic doses can only be done on a case-by-case basis when a causal relationship is established or refuted.

If appropriate information is available it is important to try to determine whether developmental toxicity is due to a specific maternally mediated mechanism or to a non-specific secondary mechanism, like maternal stress and the disruption of homeostasis. Generally, the presence of maternal toxicity should not be used to negate findings of embryo/fetal effects, unless it can be clearly demonstrated that the effects are secondary non-specific effects. This is especially the case when the effects in the offspring are significant, e.g. irreversible effects such as structural malformations. In some situations it is reasonable to assume that reproductive toxicity is due to a secondary consequence of maternal toxicity and discount the effects, for example if the chemical is so toxic that dams fail to thrive and there is severe inanition; they are incapable of nursing pups; or they are prostrate or dying.

Hazard Categories

For the purpose of classification for reproductive toxicity, chemical substances are allocated to one of two classes. Effects on reproductive ability or capacity, and on development, are considered as separate issues.

Category 1:

Known or Presumed Human Reproductive or Developmental Toxicant

This Category includes substances which are known to have produced an adverse effect on reproductive ability or capacity or on development in humans or for which there from animal studies, possibly supplemented with, to provide a strong presumption that the substance has the capacity to interfere with reproduction in humans. For regulatory purposes, a substance can be further distinguished on the basis of whether the evidence for classification is primarily from human data (Category 1A) or from animal data (Category 1B).

Category 1A to have produced an adverse effect on reproductive ability or capacity or on development in humans. The placing of the substance in this category is largely based on evidence from humans.

Category 1B to produce an adverse effect on reproductive ability or capacity or on development in humans. The placing of the substance in this category is largely based on evidence from experimental animals. Data from animal studies should provide clear evidence of specific reproductive toxicity in the absence of other toxic effects, or if occurring together with other toxic effects the adverse effect on reproduction is considered not to be a secondary non-specific consequence of other toxic effects. However, when there is mechanistic information that raises doubt about the relevance of the effect for humans, classification in Category 2 may be more appropriate.

Category 2:

Suspected human reproductive or developmental toxicant

This Category includes substances for which there is some evidence from humans or experimental animals, - possibly supplemented with other information - of an adverse effect on reproductive ability or capacity, or on development, in the absence of other toxic effects, or if occurring together with other toxic effects the adverse effect on reproduction is considered not to be a secondary non-specific consequence of the other toxic effects, and where the evidence is not sufficiently convincing to place the substance in Category 1. For instance, deficiencies in the study may make the quality of evidence less convincing, and in view of this Category 2 could be the more appropriate classification.

Effects on or Via Lactation

Effects on or via lactation are allocated to a separate single category. It is appreciated that for many substances there is no information on the potential to cause adverse effects on the offspring via lactation. However, for substances which are absorbed by women and have been shown to interfere with lactation or which may be present (including metabolites) in breast milk in amounts sufficient to cause concern for the health of a breastfed child, should be classified to indicate this property hazardous to breastfed babies. This classification can be assigned on the basis of:

a. Metabolism, distribution and excretion studies that would indicate the likelihood the substance would be present in potentially toxic levels in breast milk;

b. Of one or two generation studies in animals which provide clear evidence of adverse effect in the offspring due to transfer in the milk or adverse effect on the quality of the milk;

c. Evidence indicating a hazard to babies during the lactation period.

Basis of Classification

Classification is made on the basis of the appropriate criteria, outlined above, and an assessment of the total weight of evidence. Classification as a reproductive or developmental toxicant is intended to be used for chemicals which have an intrinsic, specific property to produce an adverse effect on reproduction or development and chemicals should not be so classified if such an effect is produced solely as a non-specific secondary consequence of other toxic effects.

In the evaluation of toxic effects on the developing offspring, it is important to consider the possible influence of maternal toxicity.

For human evidence to provide the primary basis for a Category 1A classification there must be reliable evidence of adverse effect on reproduction in humans. Evidence used for classification should ideally be from well conducted epidemiological studies which include the use of appropriate controls, balanced assessment, and due consideration of bias or confounding factors. Less rigorous data from studies in humans should be supplemented with adequate data from studies in experimental animals and classification in Category 1B should be considered.

Data already generated for classifying chemicals under existing systems should be acceptable when reviewing these chemicals with regard to classification under the harmonized system. Further testing should not normally be necessary.

Explanatory

Maternal Toxicity

Development of the offspring throughout gestation and during the early post can be influenced by toxic effects in the mother either through non-specific mechanisms related to stress and the disruption of maternal homeostasis, or by specific maternally mediated mechanisms. So, in the interpretation of the developmental outcome to decide classification for developmental effects it is important to consider the possible influence of maternal toxicity. This is a complex issue because of uncertainties surrounding the relationship between maternal toxicity and developmental outcome. Expert judgment and a weight of evidence approach, using all available studies, should be used to determine the degree of influence that should be attributed to maternal toxicity when interpreting the criteria for classification for developmental effects. The adverse effects in the embryo/fetus should be first considered, and then maternal toxicity, along with any other factors which are likely to have influenced these effects, as weight of evidence, to help reach a conclusion about classification.

Based on pragmatic observation, it is believed, that maternal toxicity may, depending on severity, influence development via non-specific secondary mechanisms, producing effects such as depressed fetal weight, retarded ossification, and possibly resorptions and certain malformations in some strains of certain species. However, the limited numbers of studies which have investigated the relationship between developmental effects and general maternal toxicity have failed to demonstrate a consistent, reproducible relationship across species. Developmental effects which occur even in the presence of maternal toxicity are considered to be evidence of developmental toxicity, unless it can be unequivocally demonstrated on a case by case basis that the developmental effects are secondary to maternal toxicity. Moreover, classification should be considered where there is significant toxic effect in the offspring, e.g. irreversible effects such as structural malformations, embryo/fetal lethality, significant post-natal functional deficiencies.

Classification should not automatically be discounted for chemicals that produce developmental toxicity only in association with maternal toxicity, even if a specific maternally mediated mechanism has been demonstrated. In such a case, classification in Category 2 may be considered more appropriate than Category 1. However, when a chemical is so toxic that maternal death or severe inanition results, or the dams are prostrate and incapable of nursing the pups, it may be reasonable to assume that developmental toxicity is produced solely as a secondary consequence of maternal toxicity and discount the developmental effects. Classification may not necessarily be the outcome in the case of minor developmental changes e.g. small reduction in fetal/pup body weight, retardation of ossification when seen in association with maternal toxicity.

Some of the end points used to assess maternal toxicity are provided below. Data on these end points, if available, needs to be evaluated in light of their statistical or biological significance and dose response relationship.

Maternal Mortality: An increased incidence of mortality among the treated dams over the controls should be considered evidence of maternal toxicity if the increase occurs in a dose related manner and can be attributed to the systemic toxicity of the test material.

Organ Specific Toxic Effects
201

Maternal mortality greater than 10% is considered excessive and the data for that dose level should not normally be considered for further evaluation:

- Mating Index (no. animals with seminal plugs or sperm/no. mated x 100)

- Fertility Index (no. animals with implants/no. of matings x 100)

- Gestation Length

Body Weight and Body Weight Change: Consideration of the maternal body weight change and/or adjusted (corrected) maternal body weight should be included in the evaluation of maternal toxicity whenever such data are available. The calculation of adjusted (corrected) mean maternal body weight change, which is the difference between the initial and terminal body weight minus the gravid uterine weight (or alternatively, the sum of the weights of the fetuses), may indicate whether the effect is maternal or intrauterine. In rabbits, the body weight gain may not be useful indicators of maternal toxicity because of normal fluctuations in body weight during pregnancy.

Food and Water Consumption: The observation of a significant decrease in the average food or water consumption in treated dams compared to the control group may be useful in evaluating maternal toxicity, particularly when the test material is administered in the diet or drinking water. Changes in food or water consumption should be evaluated in conjunction with maternal body weights when determining if the effects noted are reflective of maternal toxicity or more simply, unpalatability of the test material in feed or water.

Clinical evaluations (including clinical signs, markers, haematology and clinical chemistry studies) observation of increased incidence of significant clinical signs of toxicity in treated dams relative to the control group may be useful in evaluating maternal toxicity. If this is to be used as the basis for the assessment of maternal toxicity, the types, incidence, degree and duration of clinical signs should be reported in the study. Examples of frank clinical signs of maternal intoxication include: coma, prostration, hyperactivity, loss of righting reflex, ataxia, or labored breathing.

Post-mortem data incidence and severity of post-mortem findings may be indicative of maternal toxicity. This can include gross or microscopic pathological findings or organ weight data, e.g., absolute organ weight, organ-to-body weight ratio, or organ-to-brain weight ratio. When supported by findings of adverse histopathological effects in the affected organ(s), the observation of a significant change in the average weight of suspected target organ(s) of treated dams, compared to those in the control group, may be considered evidence of maternal toxicity.

Potency and Cut-off Doses

In the present scheme, the relative potency of a chemical to produce a toxic effect on reproduction is not included in the criteria for reaching a conclusion regarding classification. Nevertheless, during the development of this scheme it was suggested that cut-off dose levels should be included, in order to provide some means of assessing and categorizing the potency of chemicals for the ability to produce an adverse effect on reproduction. This concept has not been readily accepted by all member countries because of concerns that any specified cut-off level may be exceeded by human exposure levels in certain situations, e.g. inhalation of volatile solvents, the level may be inadequate in cases where humans are more sensitive than the animal model, and because of disagreements about whether or not potency is a component of hazard.

There has been interest in this concept to further consider it as a future development of the classification scheme.

Limit dose

Member countries appear to be in agreement about the concept of a limit dose, above which the production of an adverse effect may be considered to be outside the criteria which lead to classification. However, there is disagreement between members regarding the inclusion within the criteria of a specified dose as a limit dose. Some Test Guidelines specify a limit dose; other Test Guidelines qualify the limit dose with a statement that higher doses may be necessary if anticipated human exposure is sufficiently high that an adequate margin of exposure would not be achieved. Also, due to species differences in toxicokinetics, establishing a specific limit dose may not be adequate for situations where humans are more sensitive than the animal model.

In principle, adverse effects on reproduction seen only at very high dose levels in animal studies (for example doses that induce prostration, severe inappetence, excessive mortality) would not normally lead to classification, unless other information is available, e.g. toxicokinetics information indicating that humans may be more susceptible than animals, to suggest that classification is appropriate.

However, specification of the actual 'limit dose' will depend upon the test method that has been employed to provide the test results, e.g. in the OECD Test Guideline for repeated dose toxicity studies by the oral route, an upper dose of 1000 mg/kg unless expected human response indicates the need for a higher dose level, has been recommended as a limit dose.

Animal and Experimental Data

A number of internationally accepted test methods are available; these include methods for developmental toxicity testing, methods for peri- and post-natal toxicity testing and methods for one or two-generation toxicity testing.

Results obtained from Screening Tests (e.g. OECD Guidelines 421 - Reproduction/Developmental Toxicity Screening Test, and 422 - Combined Repeated Dose Toxicity Study with Reproduction/Development Toxicity Screening Test) can also be used to justify classification, although it is recognized that the quality of this evidence is less reliable than that obtained full studies.

Adverse effects or changes, seen in short- or long-term repeated dose toxicity studies, which are judged likely to impair reproductive ability or capacity and which occur in the absence of significant generalized toxicity, may be used as a basis for classification, e.g. histopathological changes in the gonads.

Evidence from in vitro assays, or non-mammalian tests, and from analogous substances using structure-activity relationship (SAR), can contribute to the procedure for classification. In all cases of this nature, expert judgment must be used to assess the adequacy of the data. Inadequate data should not be used as a primary support for classification.

It is preferable that animal studies are conducted using appropriate routes of administration which relate to the potential route of human exposure. However, in practice, reproductive toxicity studies are commonly conducted using the oral route, and such studies will normally be suitable for evaluating the hazardous properties of the substance with respect to reproductive toxicity. However, if it can be conclusively demonstrated that the clearly identified mechanism or mode of action has no relevance for humans or when the toxicokinetic differences are so marked that it is certain that the hazardous property will not be expressed in humans then a substance which produces an adverse effect on reproduction in experimental animals should not be classified.

Studies involving routes of administration such as intravenous or intraperitoneal injection, which may result in exposure of the reproductive organs to unrealistically high levels of the test substance, or elicit local damage to the reproductive organs, e.g. by irritation, must be interpreted with extreme caution and on their own would not normally be the basis for classification.

Classification Criteria for Mixtures

Classification of mixtures will be based on the available test data of the individual constituents of the mixture using cut-off values/concentration limits for the components of the mixture. The classification may be modified on a case-by case basis based on the available test data for the mixture as a whole. In such cases, the test results for the mixture as a whole must be shown to be conclusive taking into account dose and other factors such as duration, observations and analysis (e.g., statistical analysis, test sensitivity) of reproduction test systems. Adequate documentation supporting the classification should be retained and made available for review upon request.

Bridging Principles

Where the mixture itself has not been tested to determine its reproductive toxicity, but there are sufficient data on the individual ingredients and similar tested mixtures to adequately characterise the hazards of the mixture, this data will be used in accordance with the following agreed bridging rules. This ensures that the classification process uses the available data to the greatest extent possible in characterising the hazards of the mixture without the necessity for additional testing in animals.

Dilution

If a mixture is diluted with a diluent which is not expected to affect the reproductive toxicity of other ingredients, then the new mixture may be classified as equivalent to the original mixture.

Batching

The reproductive toxicity potential of one production batch of a complex mixture can be assumed to be substantially equivalent to that of another production batch of the same commercial product produced by and under the control of the same manufacture unless there is reason to believe there is significant variation in composition such that the reproductive toxicity potential of the batch has changed. If the latter occurs, a new classification is necessary.

Given the following:

1. Two mixtures

   i)   A + B

   ii) C + B

2. The concentration of Ingredient of Ingredient B, toxic to reproduction, is the same in both mixtures.

3. The concentration of ingredient A in mixture i equals that of ingredient C in mixture ii.

4. Data on toxicity for A and C are available and substantially equivalent, i.e. they are not expected to affect the reproductive toxicity of B.

If mixture (i) is already classified by testing, mixture (ii) can be assigned the same category.

Classification of Mixtures When Data are Available for All Components or Only for Some Components of the Mixture.

The mixture will be classified as a reproductive toxin when at least one ingredient has been classified as a Category 1 or Category 2 reproductive toxicant and is present at or above the appropriate cut-off value/concentration limit as mentioned in table below for Category 1 and 2 respectively.

Table: Cut-off values/concentration limits of ingredients of a mixture classified as reproductive toxicants that would trigger classification of the mixture.

Ingredient Classified as:	Cut-off/concentration limits triggering classification of a mixture as:	
	Category 1 reproductive toxicant	Category 2 reproductive toxicant
Category 1 reproductive toxicant	$\geq$ 0.1 % (note 1)	
	$\geq$ 0.3 % (note 2)	
Category 2 reproductive toxicant		$\geq$ 0.1 % (note 3)
		$\geq$ 3.0 % (note 4)

- If a Category 1 reproductive toxicant is present in the mixture as an ingredient at a concentration between 0.1% and 0.3%, every regulatory authority would require information on the MSDS for a product. However, a label warning would be optional. Some authorities will choose to label when the ingredient is present in the mixture between 0.1% and 0.3%, whereas others would normally not require a label in this case.

- If a Category 1 reproductive toxicant reproductive toxicant is present in the mixture as an ingredient at a concentration of $\geq$ 0.3%, both an MSDS and a label would generally be expected.

- If a Category 2 reproductive toxicant is present in the mixture as an ingredient at a concentration between 0.1% and 3.0%, every regulatory authority would require information on the MSDS for a product. However, a label warning would be optional. Some authorities will choose to label when the ingredient is present in the mixture between 0.1% and 3.0%, whereas others would normally not require a label in this case.

- If a Category 2 reproductive toxicant is present in the mixture as an ingredient at a concentration of $\geq$ 3.0%, both an MSDS and a label would generally be expected.

Hazard Communication

Allocation of Label Elements

General and specific considerations concerning labeling requirements are Annex 5 contains examples of precautionary statements and pictograms which can be used where allowed by the competent authority. Additional reference sources providing advice on the use of precautionary information is also included.

Table: Label elements for Reproductive Toxicity.

	Category 1A	Category 1B	Category 2	Additional Category
Symbol	New health hazard symbol	New health hazard symbol	New health hazard symbol	
Signal Word	Danger	Danger	Warning	
Hazard Statement	May damage fertility or the unborn child (state specific effect if known or route of exposure if it is conclusively proven that no other routes of exposure cause the hazard)	May damage fertility or the unborn child (state specific effect if known or route of exposure if it is conclusively proven that no other routes of exposure cause the hazard)	Suspected of damaging fertility or the unborn child (state specific effect if known or route of exposure if it is conclusively proven that no other routes of exposure cause the hazard)	May cause harm to breast-fed children.

Respiratory Toxicity

For every breath we take, air passes through the mouth and nasal cavity and then through the trachea, bronchus, and bronchioles before entering lung, where gas exchange occurs. However, many inhaled toxins bypass these defense mechanisms of lung, or become activated by the metabolic enzymes, leading to pulmonary toxicity and acute injury to the lung epithelia. Proper identification of potential respiratory toxins and irritants is important to numerous industries, including chemical, personal care, and pharmaceutical industries.

Many respiratory toxicity identification methods have been established in the past, such as human research, and animal study. However, pulmonary research in humans is obviously limited due to the human and ethics issues; and for whole-animal inhalation studies, significant cost and time considerations are associated.

Nowadays, an improvement has been made by using three-dimensional (3D) human derived respiratory tissue model to predict the respiratory toxicity of compounds and chemicals. These tissues exhibits human relevant tissue structure and cellular morphology with high uniformity and reproducibility, and allows for *in vitro* respiratory toxicity testing of compounds such as pathogens, chemicals or therapeutics.

Respiratory toxicity can include a variety of acute and chronic pulmonary conditions, including local irritation, bronchitis, pulmonary edema, emphysema, and cancer. It is well known that exposure to environmental and industrial chemicals can impair respiratory function. Ground-level ozone, the main component in smog, causes breathing problems, aggravates asthma, and increases the severity and incidence of respiratory infections. Acute exposure to respiratory toxicants can trigger effects ranging from mild irritation to death by asphyxiation. Prolonged exposure to respiratory toxicants can cause structural damage to the lungs, resulting in chronic diseases such as pulmonary fibrosis, emphysema, and cancer. Pulmonary fibrosis is a serious lung disease in which airways become restricted or inflamed, leading to difficulty in breathing. It can be caused by exposure to coal dust, aluminum, beryllium, and carbides of tungsten. Emphysema, a degenerative and potentially fatal disease, is characterized by the inability of the lungs to fully expand and contract. The most common cause of emphysema is heavy cigarette smoking, but the disease can also be induced by exposure to aluminum, cadmium oxide, ozone, and nitrogen oxides. In addition, several toxicants are known to cause respiratory cancer. Examples of well-established human lung carcinogens are cigarette smoke, asbestos, arsenic, and nickel.

References

- Dermal-toxicity, pharmacology-toxicology-and-pharmaceutical-science: sciencedirect.com, Retrieved 11 June 2018

- Ototoxicity, diseases: myvmc.com, Retrieved 30 March 2018

- Hepatotoxicity-due-to-drugs: medindia.net, Retrieved 13 May 2018

- Nephrotoxicity, pharmacology-toxicology-and-pharmaceutical-science: sciencedirect.com, Retrieved 21 March 2018

- What-is-Neurotoxicity: news-medical.net, Retrieved 25 June 2018

- Respiratory-toxicity: creative-bioarray.com, Retrieved 11 July 2018

Chapter 8

Other Significant Aspects of Toxicity

The complete understanding of toxicity requires a study of acute toxicity, toxicity class, lethal dose, fixed dose procedure, bioaccumulation, etc. These have been extensively elaborated in this chapter.

Acceptable Daily Intake

Acceptable daily intake is the estimate of the amount of a toxic substance in food or drinking water, expressed on a body mass basis (usually mg/kg body weight), which can be ingested daily over a lifetime by humans without appreciable health risk. For calculation of the daily intake per person, a standard body mass of 60 kg is used. The acceptable daily intake is normally used for food additives (tolerable daily intake is used for contaminants).

Purpose of an ADI

ADIs serve to protect the health of consumers and to make international trade in food easier. The ADI is a practical approach to determining the safety of food additives and is a means of achieving some harmonization of regulatory control. The advantage of regulatory and advisory bodies setting ADIs for food additives is that they are universally applicable in different countries and to all sectors of the population.

Determination of the ADI

Basically, expert scientific committees advise national and international regulatory authorities. The safety assessments of food additives have developed along similar lines in individual Member States in the European Union and in the wider international community. The main international body that addresses the safety of food additives is the Joint Expert Committee on Food Additives (JECFA) of the United Nations Food and Agriculture Organisation (FAO) and the World Health Organisation (WHO). The setting of international standards has become increasingly important in recent years as the World Trade Organisation arrangements specify that Joint FAO/WHO and Codex

Alimentarius Commission (Codex) standards apply to the safety and composition of foods worldwide. The "Codex General Standard for Food Additives" (GSFA, Codex STAN 192-1995) which was originally adopted in 1995 is currently under development and is regularly updated to include additional food additive provisions adopted by the Codex Alimentarius Commission. It outlines the conditions under which permitted food additives may be used in all foods. At EU level, additives approved for use are specified in European legislation and are given an E-number. These additives have been evaluated by the former Scientific Committee on Food (SCF) and since the creation of the European Food Safety Authority (EFSA) by their Panel on Food Additives and Nutrient Sources Added to Food (ANS). As part of its safety evaluations EFSA establishes, when possible (i.e. when sufficient information is available), an ADI for each additive.

Method to Determine the ADI

The general criteria for the use of food additives set out in the EU Directives stipulate that additives can be approved only if they present no hazard to human health at the level of use proposed based on the scientific evidence available. The safety evaluation is based on a scientific review of all pertinent toxicological data on the specific additive-both observations in humans and mandatory tests in animals. In the EU, all the evidence is reviewed by the European Food Safety Authority. The toxicological tests required by the regulatory authorities include lifetime feeding studies and multigenerational studies that determine how the additive is handled by the body in order to assess any possible harmful effects of the additive or its derivatives. The starting point for establishing the ADI is the determination of the "No Observed Adverse Effect Level" (NOAEL) for the most sensitive adverse effect relevant to human health in the most sensitive species of experimental animal. The NOAEL is, therefore, the highest dietary level of an additive at which no adverse effects were observed in the studies and it is expressed in milligrams of the additive per kilogram of bodyweight per day (mg/kg bodyweight/day). The NOAEL is then divided by a safety factor, usually 100, which results in a large margin of safety.

Necessity of a Safety Margin

Firstly, the NOAEL is determined in animals, not humans. It is therefore prudent to adjust for possible differences by assuming that man is more sensitive than the most sensitive test animal. Secondly, the reliability of toxicity tests is limited by the number of animals tested. Such tests cannot represent the diversity of the human population, subgroups of which may show different sensitivities (e.g. children, the old and the infirm). Again, it is prudent to adjust for these differences.

Safety Margins Normally Used when Determining Levels of Food Additives

Traditionally, the World Health Organisation has used a safety or uncertainty factor of 100, based on a 10-fold factor to allow for differences between animals and an average human, and a 10-fold factor to allow for differences between average humans and sensitive subgroups (pregnant women, the elderly). However, this may be varied according to the characteristics of the additive, the extent of the toxicology data and the conditions of use.

Effect of the Consumption of an Additive Above the ADI on an Individual

The consumption of an additive above its ADI on a given day is not a cause for concern because the ADI has a large built-in safety factor and in practice, consumption above the ADI on one day is more than accounted for by consumption below the ADI on most other days. As mentioned, an ADI references a life time long exposure situation, and is not a reference value for a single occasion. However, if an intake figure indicates that the ADI may be regularly exceeded by certain sectors of the population, it may be necessary for the European Food Safety Authority to advise a reduction of levels in foods consistent with the amount needed to achieve its function, or to reduce the range of foods in which the additive is permitted for use. Because of the large safety margin used in setting the ADI, it is likely that an ADI for a given additive would have to be exceeded by some considerable amount for there to be any risk of harm to human health.

Monitoring of Dietary Intakes of Food Additives

The monitoring of food additives is carried out by individual Member States on advice from the European Food Safety Authority. The ADI is compared with "average" and "extreme" consumption estimates in the population as whole or in particular subgroups of the population. Provided that intakes for average and extreme consumers are within the ADI, it is unlikely that any harm will result because the ADI is based on a no-observed adverse effect level, to which a large safety margin has been applied. To ensure that consumers are not exceeding the ADI by consuming too much or too many products containing a particular additive, EU legislation requires that intake studies be carried out to assess any changes in intake patterns.

Acute Toxicity

Acute toxicity is the effect on the human body of either a single exposure or repeated multiple exposures to a compound over a short period of time. Acute toxicity relates to adverse effects that occur within 14 days of exposure. The opposite to acute toxicity is chronic toxicity, which relates to adverse effects resulting from long term exposure to a compound. OSHA and the ACGIH set regulatory limits for short term exposure to substances that produce acute toxicity.

Acute toxicity regulatory values are found in data material sheets that are associated with the substance. The method of arriving at the limit depends on the method of ingestion, whether it is oral, dermal or by inhalation. The threshold limit value time weighted average is the maximum amount of exposure over an 8 hour work day period. The short term exposure limit (STEL), is the maximum level of exposure during any 15 minute period over an 8 hour working day.

Measures of Acute Toxicity

Regulatory Values

Limits for short-term exposure, such as STELs or CVs, are defined only if there is a particular acute toxicity associated with a substance. These limits are set by the American Conference of Governmental Industrial Hygienists (ACGIH) and the Occupational Safety and Health Administration (OSHA), based on experimental data. The values set by these organizations do not always coincide exactly, and in the chemical industry it is general practice to choose the most conservative value in order to ensure the safety of employees. The values can typically be found in a material safety data sheet. There are also different values based on the method of entry of the compound (oral, dermal, or inhalation).

- Threshold limit value-time-weighted-average: The maximum concentration to which a worker can be exposed every work day (8 hours) and experience no adverse health effects.

- Short-Term Exposure Limit, STEL or Threshold limit value-short-term exposure limit, TLV-STEL: The concentration which no person should be exposed to for more than 15 minutes during an 8-hour work day.

- Ceiling value, CV or Threshold limit value-ceiling, TLV-C: The concentration which no person should ever be exposed to.

Experimental Values

- No-observed-adverse-effect level, NOAEL

- Lowest-observed-adverse-effect level, LOAEL

- Maximum tolerable concentration, MTC, LC_0; Maximum tolerable dose, MTD, LD_0

- Minimum lethal concentration, LC_{min}; Minimum lethal dose, LD_{min}

- Median lethal concentration, LC_{50}; Median lethal dose, LD_{50}; Median lethal time, LT_{50} (LT50)

- Absolute lethal concentration, LC_{100}; Absolute lethal dose, LD_{100}

The most referenced value in the chemical industry is the median lethal dose, or LD50. This is the concentration of substance which resulted in the death of 50% of test subjects (typically mice or rats) in the laboratory.

Responses and Treatments

When a person has been exposed to an acutely toxic dose of a substance, they can be treated in a number of ways in order to minimize the harmful effects. Obviously, the severity of the response is related to the severity of the toxic response exhibited. These treatment methods include (but are not limited to):

- Emergency showers used for removing irritating or hazardous chemicals from the skin.

- Emergency eye washes used for removing any irritating or hazardous chemicals from the eyes.

- Activated charcoal used to bind and remove harmful substances consumed orally. This is used as an alternative to conventional stomach pumping.

Fixed Dose Procedure

The fixed-dose procedure (FDP), proposed in 1984 by the British Toxicology Society, is a method to assess a substance's acute oral toxicity. In this procedure the test substance is given at one of the four fixed-dose levels (5, 50, 500, and 2000 mg/kg) to five male and five female rats. The objective is to identify a dose that produces clear signs of toxicity but no mortality. Depending on the results of the first test, either no further testing is needed or a higher or lower dose is tested: If mortality occurs, retesting at a lower dose level is necessary (except if the original dose chosen is 5 mg/kg). If no signs of toxicity occur at the initial dose, it is necessary to retest at a higher dose level. The results are thus interpreted in relation to animal survival and evident toxicity and it becomes possible to assign the chemical to one of the OECD classification categories.

In comparison to the older LD_{50} test developed in 1927, this procedure produces similar results while using fewer animals and causing less pain and suffering. As a result, in 1992 this test was proposed as an alternative to the LD50 test by the Organisation for Economic Co-operation and Development under OECD Test Guideline 420. However, the U.S. Food and Drug Administration has begun to approve non-animal alternatives in response to research cruelty concerns and the lack of validity/sensitivity of animal tests as they relate to humans.

Lethal Dose

In toxicology, the lethal dose (LD) is an indication of the lethal toxicity of a given substance or type of radiation. Because resistance varies from one individual to another, the "lethal dose" represents a dose (usually recorded as dose per kilogram of subject body weight) at which a given *percentage* of subjects will die. The lethal concentration is a lethal dose measurement used for gases or particulates. The LD may be based on the standard person concept, a theoretical individual that has perfectly "normal" characteristics, and thus not apply to all sub-populations.

Median Lethal Dose (LD$_{50}$)

The median lethal dose, LD$_{50}$ (abbreviation for "lethal dose, 50%"), LC$_{50}$ (lethal concentration, 50%) or LCt$_{50}$ (lethal concentration and time) of a toxin, radiation, or pathogen is the dose required to kill half the members of a tested population after a specified test duration. LD$_{50}$ figures are frequently used as a general indicator of a substance's acute toxicity. A lower LD$_{50}$ is indicative of increased toxicity.

The test was created by J.W. Trevan in 1927. The term "semilethal dose" is occasionally used with the same meaning, in particular in translations from non-English-language texts, but can also refer to a *sub*lethal dose; because of this ambiguity, it is usually avoided. LD$_{50}$ is usually determined by tests on animals such as laboratory mice. In 2011 the US Food and Drug Administration approved alternative methods to LD$_{50}$ for testing the cosmetic drug Botox without animal tests.

LD values for humans are best estimated by extrapolating results from human cell cultures. One form of measuring LD is to use animals like mice or rats, converting to dosage per kilogram of biomass, and extrapolating to human norms. The degree of error from animal-extrapolated LD values is large. The biology of test animals differs in important aspects to that of humans. For instance, mouse tissue is approximately fifty times less responsive than human tissue to the venom of the Sydney funnel-web spider. The square-cube law also complicates the scaling relationships involved. Researchers are shifting away from animal-based LD measurements in some instances. The U.S. Food and Drug Administration has begun to approve more non-animal methods in response to animal welfare concerns.

The LD$_{50}$ is usually expressed as the mass of substance administered per unit mass of test subject, typically as milligrams of substance per kilogram of body mass, but stated as nanograms (suitable for botulinum), micrograms, milligrams, or grams (suitable for paracetamol) per kilogram. Stating it this way allows the relative toxicity of different substances to be compared, and normalizes for the variation in the size of the animals exposed, although toxicity does not always scale simply with body mass.

The choice of 50% lethality as a benchmark avoids the potential for ambiguity of making measurements in the extremes and reduces the amount of testing required. However, this also means that LD$_{50}$ is *not* the lethal dose for all subjects; some may be killed by much less, while others survive doses far higher than the LD$_{50}$. Measures such as "LD$_{1}$" and "LD$_{99}$" (dosage required to kill 1% or 99%, respectively, of the test population) are occasionally used for specific purposes.

Lethal dosage often varies depending on the method of administration; for instance, many substances are less toxic when administered orally than when intravenously administered. For this reason, LD_{50} figures are often qualified with the mode of administration, e.g., "LD_{50} i.v."

The related quantities $LD_{50}/_{30}$ or $LD_{50}/_{60}$ are used to refer to a dose that without treatment will be lethal to 50% of the population within (respectively) 30 or 60 days. These measures are used more commonly with radiation, as survival beyond 60 days usually results in recovery.

Median Infective Dose

The median infective dose (ID_{50}) is the number of organisms received by a person or test animal qualified by the route of administration (e.g., 1,200 org/man per oral). Because of the difficulties in counting actual organisms in a dose, infective doses may be expressed in terms of biological assay, such as the number of LD_{50}'s to some test animal. In biological warfare infective dosage is the number of infective doses per minute for a cubic meter (e.g., ICt_{50} is 100 medium doses - min/m^3).)

Lowest Lethal Dose

The lowest lethal dose (LD_{Lo}) is the least amount of drug that can produce death in a given animal species under controlled conditions. The dosage is given per unit of bodyweight (typically stated in milligrams per kilogram) of a substance known to have resulted in fatality in a particular species. When quoting an LD_{Lo}, the particular species and method of administration (*e.g.* ingested, inhaled, intravenous) are typically stated.

Median Lethal Concentration

For gases and aerosols, lethal concentration (given in mg/m^3 or ppm, parts per million) is the analogous concept, although this also depends on the duration of exposure, which has to be included in the definition. The term incipient lethal level is used to describe a LC_{50} value that is independent of time.

A comparable measurement is LCt_{50}, which relates to lethal dosage from exposure, where C is concentration and t is time. It is often expressed in terms of $mg-min/m^3$. LCt_{50} is the dose that will cause incapacitation rather than death. These measures are commonly used to indicate the comparative efficacy of chemical warfare agents, and dosages are typically qualified by rates of breathing (e.g., resting = 10 l/min) for inhalation, or degree of clothing for skin penetration. The concept of Ct was first proposed by Fritz Haber and is sometimes referred to as Haber's Law, which assumes that exposure to 1 minute of 100 mg/m^3 is equivalent to 10 minutes of 10 mg/m^3 ($1 \times 100 = 100$, as does $10 \times 10 = 100$).

Some chemicals, such as hydrogen cyanide, are rapidly detoxified by the human body, and do not follow Haber's Law. So, in these cases, the lethal concentration may be given simply as LC_{50} and qualified by a duration of exposure (e.g., 10 minutes). The Material Safety Data Sheets for toxic substances frequently use this form of the term even if the substance does follow Haber's Law.

Lowest Lethal Concentration

The LC_{Lo} is the lowest concentration of a chemical, given over a period of time, that results in the fatality of an individual animal. LC_{Lo} is typically for an acute (<24 hour) exposure. It is related to the LC_{50}, the median lethal concentration. The LC_{Lo} is used for gases and aerosolized material.

Limitations

As a measure of toxicity, lethal dose is somewhat unreliable and results may vary greatly between testing facilities due to factors such as the genetic characteristics of the sample population, animal species tested, environmental factors and mode of administration.

There can be wide variability between species as well; what is relatively safe for rats may very well be extremely toxic for humans (*cf.* paracetamol toxicity), and vice versa. For example, chocolate, comparatively harmless to humans, is known to be toxic to many animals. When used to test venom from venomous creatures, such as snakes, LD_{50} results may be misleading due to the physiological differences between mice, rats, and humans. Many venomous snakes are specialized predators of mice, and their venom may be adapted specifically to incapacitate mice; and mongooses may be exceptionally resistant. While most mammals have a very similar physiology, LD_{50} results may or may not have equal bearing upon every mammal species, including humans.

Animal Rights Concerns

Animal-rights and animal-welfare groups, such as Animal Rights International, have campaigned against LD_{50} testing on animals in particular as, in the case of some substances, causing the animals to die slow, painful deaths. Several countries, including the UK, have taken steps to ban the oral LD_{50}, and the Organisation for Economic Co-operation and Development (OECD) abolished the requirement for the oral test in 2001.

Toxicity Class

Toxicity class refers to a classification system for pesticides that has been created by a national or international government-related or -sponsored organization. It addresses the acute toxicity of agents such as soil fumigants, fungicides, herbicides, insecticides, miticides, molluscicides, nematicides, or rodenticides.

General Considerations

Assignment to a toxicity class is based typically on results of acute toxicity studies such as the determination of LD_{50} values in animal experiments, notably rodents, via oral, inhaled, or external application. The experimental design measures the acute death rate of an agent. The toxicity class generally does not address issues of other potential harm of the agent, such as bioaccumulation, issues of carcinogenicity, teratogenicity, mutagenic effects, or the impact on reproduction.

Regulating agencies may require that packaging of the agent be labeled with a signal word, a specific warning label to indicate the level of toxicity.

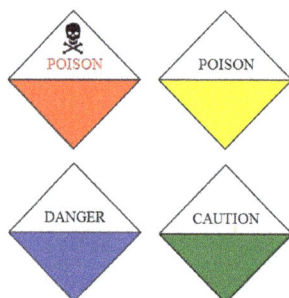

Figure: Indian Toxicity label system

Figure: Toxicity symbol for European

By Jurisdiction

World Health Organization

The World Health Organization (WHO) names four toxicity classes:

- Class I – a: Extremely hazardous
- Class I – b: Highly hazardous
- Class II: Moderately hazardous
- Class III: Slightly hazardous

The system is based on LD50 determination in rats, thus an oral solid agent with an LD50 at 5 mg or less/kg bodyweight is Class Ia, at 5–50 mg/kg is Class Ib, LD50 at 50–2000 mg/kg is Class II, and at LD50 at the concentration more than 2000 mg/kg is classified as Class III. Values may differ for liquid oral agents and dermal agents.

European Union

There are eight toxicity classes in the European Union's classification system, which is regulated by Directive 67/548/EEC:

- Class I: Very toxic
- Class II: Toxic

- Class III: Harmful
- Class IV : Corrosive
- Class V : Irritant
- Class VI : Sensitizing
- Class VII : Carcinogenic
- Class VIII : Mutagenic

Very toxic and toxic substances are marked by the European toxicity symbol.

United States

The United States Environmental Protection Agency (EPA) uses four toxicity classes in its toxicity category rating. Classes I to III are required to carry a signal word on the label. Pesticides are regulated in the United States primarily by the Federal Insecticide, Fungicide, and Rodenticide Act (FIFRA).

Toxicity Class I

- Most toxic;
- Requires signal word: "Danger-Poison", with skull and crossbones symbol, possibly followed by:

  "Fatal if swallowed", "Poisonous if inhaled", "Extremely hazardous by skin contact--rapidly absorbed through skin", or "Corrosive--causes eye damage and severe skin burns".

Class I materials are estimated to be fatal to an adult human at a dose of less than 5 grams (less than a teaspoon).

Toxicity Class II

- Moderately toxic
- Signal word: "Warning", possibly followed by:

  "Harmful or fatal if swallowed", "Harmful or fatal if absorbed through the skin", "Harmful or fatal if inhaled", or "Causes skin and eye irritation".

Class II materials are estimated to be fatal to an adult human at a dose of 5 to 30 grams.

Toxicity Class III

- Slightly toxic
- Signal word: Caution, possibly followed by:

"Harmful if swallowed", "May be harmful if absorbed through the skin", "May be harmful if inhaled", or "May irritate eyes, nose, throat, and skin".

Class III materials are estimated to be fatal to an adult human at some dose in excess of 30 grams.

Toxicity Class IV

- Practically nontoxic

- No Signal Word required since 2002

General Versus Restricted Use

Furthermore, the EPA classifies pesticides into those anybody can apply (*General Use Pesticides*), and those that must be applied by or under the supervision of a certified individual. Application of *Restricted use pesticides* requires that a record of the application be kept.

Bioaccumulation

An important process through which chemicals can affect living organisms is bioaccumulation. Bioaccumulation means an increase in the concentration of a chemical in a biological organism over time, compared to the chemical's concentration in the environment. Compounds accumulate in living things any time they are taken up and stored faster than they are broken down (metabolized) or excreted. Understanding the dynamic process of bioaccumulation is very important in protecting human beings and other organisms from the adverse effects of chemical exposure, and it has become a critical consideration in the regulation of chemicals.

A number of terms are used in conjunction with bioaccumulation. Uptake describes the entrance of a chemical into an organism -- such as by breathing, swallowing, or absorbing it through the skin -- without regard to its subsequent storage, metabolism, and excretion by that organism.

Storage, a term sometimes confused with bioaccumulation, means the temporary deposit of a chemical in body tissue or in an organ. Storage is just one facet of chemical bioaccumulation. (The term also applies to other natural processes, such as the storage of fat in hibernating animals or the storage of starch in seeds).

Bioconcentration is the specific bioaccumulation process by which the concentration of a chemical in an organism becomes higher than its concentration in the air or water around the organism. Although the process is the same for both natural and manmade chemicals, the term bio-concentration usually refers to chemicals foreign to the organism. For fish and other aquatic animals, bioconcentration after uptake through the gills (or sometimes the skin) is usually the most important bioaccumulation process.

Biomagnification describes a process that results in the accumulation of a chemical in an organism at higher levels than are found in its food. It occurs when a chemical becomes more and more concentrated as it moves up through a food chain -- the dietary linkages between single-celled plants and increasingly larger animal species.

A typical food chain includes algae eaten by the water flea eaten by a minnow eaten by a trout and finally consumed by an osprey (or human being). If each step results in increased bioaccumulation, that is, biomagnification, then an animal at the top of the food chain, through its regular diet, may accumulate a much greater concentration of chemical than was present in organisms lower in the food chain.

Biomagnification is illustrated by a study of DDT which showed that where soil levels were 10 parts per million (ppm), DDT reached a concentration of 141 ppm in earthworms and 444 ppm in robins. Through biomagnification, the concentration of a chemical in the animal at the top of the food chain may be high enough to cause death or adverse effects on behavior, reproduction, or disease resistance and thus endanger that species, even when levels in the water, air, or soil are low. Fortunately, bioaccumulation does not always result in biomagnification.

The Bioaccumulation Process

Bioaccumulation is a normal and essential process for the growth and nurturing of organisms. All animals, including humans, daily bioaccumulate many vital nutrients, such as vitamins A,D and K, trace minerals, and essential fats and amino acids. What concerns toxicologists is the bioaccumulation of substances to levels in the body that can cause harm. Because bioaccumulation is the net result of the interaction of uptake, storage and elimination of a chemical, these parts of the process will be examined further.

Uptake

Bioaccumulation begins when a chemical passes from the environment into an organism's cells. Uptake is a complex process which is still not fully understood. Scientists have learned that chemicals tend to move, or diffuse, passively from a place of high concentration to one of low concentration. The force or pressure for diffusion is called the chemical potential, and it works to move a chemical from outside to inside an organism.

A number of factors may increase the chemical potential of certain substances. For example, some chemicals do not mix well with water. They are called lipophilic, meaning "fat loving," or hydrophobic, meaning "water hating." In either case, they tend to move out of water and enter the cells of an organism, where there are lipophilic microenvironments.

Storage

The same factors affecting the uptake of a chemical continue to operate inside an organism, hindering a chemical's return to the outer environment. Some chemicals are attracted to certain sites, and by binding to proteins or dissolving in fats, they are temporarily stored. If uptake slows or is not continued, or if the chemical is not very tightly bound in the cell, the body can eventually eliminate the chemical.

One factor important in uptake and storage is water solubility; the ability of a chemical to dissolve in water. Usually, compounds that are highly water soluble have a low potential to bioaccumulate and do not leave water readily to enter the cells of an organism. Once inside, they are easily removed unless the cells have a specific mechanism for retaining them.

Heavy metals like mercury and certain other water-soluble chemicals are such an exception, because they bind tightly to specific sites within the body. When binding occurs, even highly water-soluble chemicals can accumulate. This is illustrated by cobalt, which binds very tightly and specifically to sites in the liver and accumulate there despite its water solubility. Similar accumulation processes occur for mercury, copper, cadmium, and lead.

Many fat-loving (lipophilic) chemicals pass into organism's cells through the fatty layer of cell membranes more easily than water-soluble chemicals. Once inside the organism, these chemicals may move through numerous membranes until they are stored in fatty tissues and begin to accumulate.

The storage of toxic chemicals in fat reserves serves to detoxify the chemical, or at least removes it from harms way. However, when fat reserves are called upon to provide energy for an organism the materials stored in the fat may be remobilized within the organism and may again be potentially toxic. If appreciable amounts of a toxin are stored in fat and fat reserves are quickly used, significant toxic effects may be seen from the remobilization of the chemical.

Elimination

Another factor affecting bioaccumulation is whether an organism can break down and excrete a chemical. The biological breakdown of chemicals is termed metabolism. This ability varies among individual organisms and species and also depends on characteristics of the chemical itself.

Chemicals that dissolve readily in fat but not in water tend to be more slowly eliminated by the body and thus have a greater potential to accumulate. Many metabolic reactions change a chemical into more water soluble forms called metabolites, that are readily excreted.

There are exceptions, however. Natural pyrethrins, insecticides that are derived from the chrysanthemum plant, are highly fat-soluble pesticides, but they are easily degraded and do not accumulate. The insecticide chlorpyrifos, which is less fat-soluble but more poorly degraded, tends to bioaccumulate. Factors affecting metabolism often determine whether a chemical achieves its bioaccumulation potential in a given organism.

Bioaccumulation: State of Dynamic Equilibrium

When a chemical enters the cells of an organism, it is distributed and then excreted, stored or metabolized. Excretion, storage, and metabolism decrease the concentration of the chemical inside the organism, increasing the potential of the chemical in the outer environment to move into the organism. During constant environmental exposure to a chemical, the amount of a chemical accumulated inside the organism, and the amount leaving, reach a state of dynamic equilibrium.

To understand this concept of dynamic equilibrium, imagine a tub filling with water from a faucet at the top and draining out through a pipe of smaller size at the bottom. When the water level in the tub is low, little pressure is exerted on the outflow at the bottom of the tub. As the water level rises, the pressure on the outflow increases. Eventually, the amount of the water flowing out will equal the amount flowing in, and the level of the tub will not change. If the input or outflow is changed, the water in the tub adjusts to a different level.

It is the same concept with living organisms. An environmental chemical will at first move into an organism more rapidly than it is stored, degraded, and excreted. With constant exposure, its concentration inside the organism gradually increases. Eventually, the concentration of the chemical inside the organism will reach an equilibrium with the concentration of the chemical outside the organism, and the amount of chemical entering the organism will be the same as the amount leaving. Although the amount inside the organism remains constant, the chemical continues to be taken up, stored, degraded, and excreted.

If the environmental concentration of the chemical increases, the amount inside the organism will increase until it reaches a new equilibrium. Exposure to large amounts of a chemical for a long period of time, however, may overwhelm the equilibrium (for example, overflowing the tub) potentially causing harmful effects.

Likewise, if the concentration in the environment decreases, the amount inside the organism will also decline. Should the organism move to a clean environment, so that exposure ceases, then the chemical eventually will be eliminated from the body.

Factors Affecting Bioaccumulation

This simplified explanation does not take into account all of the many factors that affect the ability of chemicals to be bioaccumulated. Some chemicals bind to specific sites in the body, prolonging their stay, whereas others move freely in and out. The time between uptake and eventual elimination of a chemical directly affects bioaccumulation. Chemicals that are immediately eliminated, for example, do not bioaccumulate.

Similarly, the duration of exposure is also a factor in bioaccumulation. Most exposures to chemicals in the environment vary continually in concentration and duration, sometimes including periods of no exposure. In these cases, equilibrium is never achieved and the accumulation is less than expected.

Bioaccumulation varies between individual organisms as well as between species. Large, fat, long-lived individuals or species with low rates of metabolism or excretion of a chemical will bioaccumulate more than small, thin, short-lived organisms. Thus, an old lake trout may bioaccumulate much more than a young bluegill in the same lake.

References

- Schultz, Eric (2013). Fish Physiology: Euryhaline Fishes, Volume 32. Academic Press. pp. 481–482. ISBN 978-0-12-396951-4

- Qas-on-acceptable-daily-intakes-adis, understanding-science: eufic.org, Retrieved 15 April 2018

- Encyclopedia of Genetics, Genomics, Proteomics and Informatics. Springer. 2008. doi:10.1007/978-1-4020-6754-9_9257. ISBN 978-1-4020-6754-9

- "The MSDS HyperGlossary: Acute toxicity". Safety Emporium. Archived from the original on 16 October 2006. Retrieved 2006-11-15

- Bioaccum: extoxnet.orst.edu: Retrieved 20 March 2018

- Encyclopedia of Genetics, Genomics, Proteomics and Informatics. Springer. 2008. doi:10.1007/978-1-4020-6754-9_9257. ISBN 978-1-4020-6754-9

- Cutler, David (February 1, 2001). "Death of LD50". Trends in Pharmacological Sciences. 22. doi:10.1016/S0165-6147(00)01627-8

- Acute-toxicity-4798: safeopedia.com, Retrieved 12 June 2018

- "In U.S., Few Alternatives To Testing On Animals". Washington Post. 12 April 2008. Archived from the original on 12 November 2012. Retrieved 2011-06-26

Permissions

All chapters in this book are published with permission under the Creative Commons Attribution Share Alike License or equivalent. Every chapter published in this book has been scrutinized by our experts. Their significance has been extensively debated. The topics covered herein carry significant information for a comprehensive understanding. They may even be implemented as practical applications or may be referred to as a beginning point for further studies.

We would like to thank the editorial team for lending their expertise to make the book truly unique. They have played a crucial role in the development of this book. Without their invaluable contributions this book wouldn't have been possible. They have made vital efforts to compile up to date information on the varied aspects of this subject to make this book a valuable addition to the collection of many professionals and students.

This book was conceptualized with the vision of imparting up-to-date and integrated information in this field. To ensure the same, a matchless editorial board was set up. Every individual on the board went through rigorous rounds of assessment to prove their worth. After which they invested a large part of their time researching and compiling the most relevant data for our readers.

The editorial board has been involved in producing this book since its inception. They have spent rigorous hours researching and exploring the diverse topics which have resulted in the successful publishing of this book. They have passed on their knowledge of decades through this book. To expedite this challenging task, the publisher supported the team at every step. A small team of assistant editors was also appointed to further simplify the editing procedure and attain best results for the readers.

Apart from the editorial board, the designing team has also invested a significant amount of their time in understanding the subject and creating the most relevant covers. They scrutinized every image to scout for the most suitable representation of the subject and create an appropriate cover for the book.

The publishing team has been an ardent support to the editorial, designing and production team. Their endless efforts to recruit the best for this project, has resulted in the accomplishment of this book. They are a veteran in the field of academics and their pool of knowledge is as vast as their experience in printing. Their expertise and guidance has proved useful at every step. Their uncompromising quality standards have made this book an exceptional effort. Their encouragement from time to time has been an inspiration for everyone.

The publisher and the editorial board hope that this book will prove to be a valuable piece of knowledge for students, practitioners and scholars across the globe.

Index

www.ingramcontent.com/pod-product-compliance
Lightning Source LLC
Chambersburg PA
CBHW061950190326
41458CB00009B/2840